W9-ARY-438

Scientists at Work

Profiles of Today's Groundbreaking Scientists From

Science Times

The New York Times

Edited by Laura Chang

Foreword by Stephen Jay Gould

Introduction by Cornelia Dean

McGraw-Hill

New York   San Francisco   Washington, D.C.   Auckland   Bogotá
Caracas   Lisbon   London   Madrid   Mexico City   Milan
Montreal   New Delhi   San Juan   Singapore
Sydney   Tokyo   Toronto

McGraw-Hill

A Division of The **McGraw·Hill** Companies

1 2 3 4 5 6 7 8 9 0 DOC/DOC 0 9 8 7 6 5 4 3 2 1 0

ISBN 0-07-135882-X

Printed and bound by R. R. Donnelley & Sons Company.

McGraw-Hill books are available at special quantity discounts to use as premiums and sales promotions, or for use in corporate training programs. For more information, please write to the Director of Special Sales, Professional Publishing, McGraw-Hill, Two Penn Plaza, New York, NY 10121-2298.

 This book is printed on recycled, acid-free paper containing a minimum of 50% recycled, de-inked fiber.

Contents

Foreword

Stephen Jay Gould

We all recognize the Hollywood B-movie and pulp novel cliché of the scientist as dehumanized, deranged, dedicated only to his own vision or invention, and morally unfit. However, in our worthy efforts to debunk such nonsense, we have often touted the opposite, if benign, cliché of successful scientists as "just plain folks," sharing no traits in common to mark their distinctive achievements, but spanning the full range of human propensities and personalities. The enormous diversity of aims and types among the people described in this book might seem, at first, to support this assessment of "no different" or "nothing special." But we may, I think, defend two worthy parts of the claim for nondistinctness, while still acknowledging the special characteristics of these men and women with a different argument that removes the stigma of elitism usually associated with such assertions.

As the first source of diversity, science itself spans such a range of activities that almost any mental or technical skill can find useful employment in some room among the many mansions of the full enterprise. Science, in its broadest charge, tries to ascertain the factual reality of the natural world, and to propose and test reasons why natural phenomena operate as they do, rather than in some other conceivable but unrealized way. These two components of fact and theory call upon all varieties of human capability, from the tenacity, patience, and love of observational detail in the traditional field naturalist (see Terry DeBruyn on "over 1,000 hours walking with bears," Jim Hammill on wolves, P. Kirk Visscher on bees, and Kathy Schick and Nicholas Toth's mixture of experiment and observation in trying to understand how our earliest tool-using ancestors fashioned and employed their implements), to the disembodied abstractions of pure mathematics that seem so divorced from the world's "stuff," but that have so often led us to understand how the stuff works (John Conway memorizing 1,000 places of pi for pure fun and mental discipline; Andrew Wiles's solution of Fermat's last theorem. And lest we fall into the stereotype of mathematics as purely personal and idiosyncratic genius, we also learn that Wiles built his proof upon 300 years of partial solutions and revealing clues, thus illustrating the social and incremental side of science, even in areas least suited to solving problems by factual accumulation).

The second source of diversity must reach outside science to praise the virtues of democracy. Only a few males of wealth and breeding enjoyed any real prospect of a successful scientific career before the last few generations. But a high percentage of the people profiled in this book (including the author of this foreword) grew up in poor families with no previous traditions or chances for higher education, and gained their precious opportunity from public schools, civic libraries, and dedicated mentors. Others have transcended the even higher barriers imposed by externalities of more stringent upbringings in poor schools of the developing world, or by internalities that we usually call "handicaps," but that may provide a different way of seeing, well suited to an important room in the mansions of science (my dear friend Gary Vermeij, a blind naturalist who has made significant discoveries about the history of shells by feeling properties that his sighted colleagues—including myself—never saw, and Temple Grandin, an autistic biologist who has devised gentler ways of treating animals slaughtered for human consumption).

Nonetheless, underlying all this enormous diversity, we do detect common traits that probably unite people of unusual accomplishment in all human endeavors, and not only in science. Two of the three major components of success receive general acknowledgment. One must, first of all, possess the unusual level of mental skill required for high achievement (the equivalent of bodily gifts as a prerequisite to stardom in athletics. Fortunately, science encompasses a vast range of different skills, and almost everyone will possess a special talent for some point in the array). Secondly, despite all heroic tales of obstacles overcome, not all barriers can be transcended. Even the most glowing genius could not have understood the solar system before the invention of the telescope; even the finest intellect may be destroyed by ravages of mental or physical illness.

But beyond these two components of internal mentality and external opportunity, the third requirement of a temperament suited to such achievement rarely receives its due. (Mental skills bear little if any correlation with these requisite gifts of temperament, so I do regard these two largely internal factors of success as effectively independent. Many geniuses end up in the gutter, or in jail.) In fact, as I survey the few people of genius that I have been privileged to know, I find least variation in these temperamental features. These rare people encompass a maximal variety of hypertrophied mental skills and social backgrounds, but they all share a few precious, and preciously rare, traits of temperament that allow inborn skills to flourish and impart the tenacity to overcome external obstacles.

I find the same unity of temperament among the people profiled in this book: a mixture of confidence, optimism, stubbornness, questioning, and the simple bodily energy (combined with the mental drive) to work all the time, every single day (see profiles of Anthony Fauci and Judith Lea Swain for particular emphasis on this basic trait of living larger than life). Untold people who do not end up in such books possess the way in far greater abundance, but lack the will. If we had to summarize this elusive requirement of temperament, we could do no better than to cite a single word encompassing both the external fortune and the internal basis of success. All these people hold abundant enthusiasm, which literally means "the intake of God." They all maintain, however long they live, the childlike wonder that transcends all fear of ridicule, and that relishes the unknown or unknowable. In short, they would all dare, in their own particular way, to pose P. Kirk Visscher's closing question: "I wonder what it is to be a bee?"

Acknowledgments

Laura Chang

As a selection of profiles that appeared in *Science Times* from 1993 to 2000, this book stands upon the work of many people: Cornelia Dean, science editor of *The New York Times,* who played a large role in the early stages of this book; Nicholas Wade, her predecessor; and the editors on their staffs, Dennis Overbye, James Gorman, William Dicke, C. Claiborne Ray, Barbara Strauch, John Wilson and Linda Villarosa. A bow should also be taken by Mitchel Levitas, who supervised development of the book; the writers of the original profiles and the new pieces that accompany them; John Forbes, the science picture editor; Amy Murphy and Griffin Hansbury of McGraw-Hill; Stephanie S. Landis at North Market Street Graphics; Fred Norgaard and Farhana Hossain, for help with art; Erica Goode, Mark Hokoda, Robert A. Saar and Archie Tse for sage advice; and, of course, the scientists here, whose lives and thoughts intrigue and inspire.

This book was funded in part by a grant from the Alfred P. Sloan Foundation to enhance public understanding of science through the lives and work of individual scientists.

Introduction

Cornelia Dean

If you write about science for a newspaper like *The New York Times,* it is easy to start to think about science as a series of findings, as a stream of results reported from this laboratory or that research institute and described in the scientific journals that land in our mailboxes every day.

But science is not an enterprise of discrete findings. It is a process full of false starts, blind alleys and misguided efforts. In a sense, it is a quest. It is a search for truth. It is only when researchers prevail over their own ignorance, when they persevere through the frustration, the exhaustion and the tedium of the laboratory bench or the field, that the results that excite the journalists begin to shine through.

It was to describe this process that the reporters of *The Times* began to spend a bit more of their time talking to scientists and visiting them in their labs or at their research sites with an eye to describing scientists at work.

The approach has brought several advantages to our readers. For one thing, we're all interested in people. For most of us, the drama of a scientist's quest is far more engaging than the bland data of a research report. For another thing, most of the scientists we write about have achieved some prominence in their fields—often by unconventional methods.

So our readers have learned about scientists who figured out paleolithic tool-making techniques by teaching themselves how to make stone tools. Or researchers who stepped into the public spotlight to plead for funds for AIDS research or exploration in space. And they have read about scientists who paid a heavy price for their scientific obsessions, from the cancer researcher who endured decades of scorn until his theories began to pan out, to the biologist who suffered nerve damage inflicted by the vicious microbe she was studying, to the cosmologist who plowed ahead with his research into the fate of the universe even as disease inexorably restricted his own movements on Earth.

At least as important, reading about scientists at work offers people an avenue into the work itself. In the hands of a skillful reporter, the scientists' explanations of their research, in their own words, become avenues for understanding sometimes difficult topics, be they the complexities of mathematical theory or the physics of the subatomic world or the structure of ant societies.

Finally, the stereotypical science nerd notwithstanding, high-achieving scientists are usually passionate, interesting people, ideal grist for the writer's mill. It is fascinating to discover a biologist whose major sideline is the anatomy of extraterrestrials in *The X-Files,* a patent lawyer whose novel theory about the origin of life turned the heads of experts or a chemist who speaks the language of poetry.

Some people profiled here are giants in their fields. Others flashed briefly across the scientific firmament and then retired to obscurity. Some had scientific careers that, despite impressive early achievement, hit a dead end—like the biochemist who won the Nobel Prize and then gave it all up for the life of a surfer. And sometimes they are people like the cartoonist Gary Larson, who has never been in the thick of scientific research, but whose work is probably displayed in more laboratories than anyone else's.

The 50 profiles that follow were published in the weekly *Science Times* section of *The Times* from 1993 to 2000. Each profile is accompanied by a short sidebar, written expressly for this book, that amplifies an aspect of the scientist's life, discusses recent activities or lays out an issue provoked by the subject's research.

The editor of this book, Assistant Science Editor Laura Chang, has not attempted to group the scientists here by subject. Many have such wide-ranging interests it would be difficult to place them firmly in only one field! Instead, Laura has arranged the profiles to mix up the "hard science" and "soft tissue" people, the icons and the one-hit wonders, the generalists and obsessives, not to mention the men and women.

Like our coverage in the newspaper itself, this book does not pretend to be comprehensive. Though Laura has chosen people from a variety of fields and backgrounds, it does not offer a complete portrait of science. It is not even a complete collection of our profiles. And even if it were, no group of reporters—even if they are as gifted as those on the staff of *The*

Times—can hope to paint a complete picture of modern science simply by observing scientists at work.

Instead, it is a glimpse into a fascinating world, one we hope will inspire our readers with greater interest in science, one of the greatest enterprises of the human spirit.

Cornelia Dean is science editor of *The New York Times* and author of *Against the Tide*.

Chemistry's Poet Seeks Beauty in Atoms

Courtesy of Roald Hoffmann

Roald Hoffmann—Nobel laureate, professor of chemistry at Cornell University and poet—glanced benignly at the stream of students passing his office window along a tranquil campus street. "I love chemistry," he remarked, "because it's sort of human in scale—infinitely complex, but always tangible, always real."

Although Dr. Hoffmann's pursuits include such practical goals as finding a cheap substitute for one of the precious metals used in automobile catalytic converters, he is mainly interested in the beauty that he perceives underlying even mundane applications of chemistry. Dr. Hoffmann's students know that his conception of beauty in chemistry particularly refers to the subtle symmetries and asymmetries of electron orbitals in complex molecules. He was awarded the 1981 Nobel Prize in chemistry for his idea that chemical transformations could be approximately predicted from these symmetries. His method gave chemists a powerful tool for predicting the pathways that chemical reactions may follow.

His life is driven by the urge to share this beauty with others, in the language of poetry as well as that of quantum mechanics. And so he is also the author of a series of books of essays, criticism and poetry, the latest of which, *Chemistry Imagined,* written with Vivian Torrence and published by Smithsonian Institution Press, draws scientific allusion into poetry in somewhat the way poets draw symbolic allusions from classical mythology.

The quest for beauty leads almost inevitably to interesting science, in his view. A molecule may be needed because it is useful to industry or essential to life, but the best way to sketch it, Dr. Hoffmann says, may be to follow one's aesthetic sense rather than some mechanical recipe.

Palpable though chemical science may be, Dr. Hoffmann is not the kind of chemist who stains his fingers with laboratory reagents or cajoles recalcitrant salts into crystallizing from their solutions. He is a theorist and a poet, whose chemical intuition and personalized tool kit of mathematical tricks have pointed paths through many a knotty problem that had stumped experimentalists. Dr. Hoffmann's laboratory consists of his desk and his fountain pen, with which he produces graceful calligraphy as well as little pictures and diagrams much admired by other chemists, which symbolically describe the fruits of his theoretical research. He has barred all conventional laboratory paraphernalia—even computers—from his memento-cluttered office.

It is the awareness and appreciation of the aesthetic aspects of science rather than the mere toting up of numbers that leads to discovery, Dr. Hoffmann tells his graduate and postdoctoral students. In helping to impart this sense in them, he believes, he converts them into colleagues, fellow researchers as well as students.

At a recent symposium Dr. Hoffmann conducted for 10 of his fledgling scientists, the deceptively simple assignment had been to bring anything to the meeting that seemed interesting or beautiful. At the end of two hours, the students, including one Russian, one Mexican and two Chinese, had analyzed, discussed and pondered enough material to launch a couple of doctoral dissertations, despite the group's steady stream of wisecracks over pizza and soft drinks.

"You can see why it is impossible for me to distinguish between teaching and research, much though the accountants would like me to do so," Dr. Hoffmann said later. "Science is a bridge between teaching and discovery, and my life has been spent building bridges."

Dr. Hoffmann was born Roald Safran on July 18, 1937, in Zloczow, a town in the part of Poland that was later absorbed by the Soviet Union. After the German invasion in 1939 he and his parents, Hillel and Clara Safran, were sent to a labor camp as part of the Nazi roundup of Jews. Roald and his mother escaped, but Mr. Safran remained in the camp to help organize a breakout. The plan was betrayed, Mr. Safran was killed, and his wife and son remained hidden in the attic of a schoolhouse for the rest of the war.

Having survived the war, mother and son emigrated to the United States in 1949. His mother remarried and the boy took his stepfather's surname, Hoffmann. The name Roald had been bestowed on him in honor of Roald Amundsen, the great Norwegian polar explorer.

Young Roald got such good grades in science that he was accepted at Stuyvesant High School, one of New York City's pre-eminent science schools. "The concentration of intellect was really higher at Stuyvesant than anywhere else in my experience, including Columbia College and Harvard University graduate school," he recalls.

His academic experience at Columbia nearly diverted him from a career in science. "At the time, Columbia's chemistry department was not as inspiring as the humanities departments," he said. "Mark Van Doren taught me poetry, I studied Japanese literature and I almost switched my major to art history. Like many middle-class American Jewish parents, mine wanted me to be a doctor, so I majored in premed, and the premed courses included chemistry. The chemistry fascinated me, and eventually I found the strength to declare that I would not become a doctor but would become a chemist."

But he soon discovered that he was not cut out to be a laboratory chemist. "In graduate school, one time, I was supposed to synthesize a compound—a polyphenyl porphyrin—using a sealed pressure vessel to contain the reaction. I didn't seal it properly, and the thing blew up, demolishing a fume hood and smearing gunk all over a brand-new laboratory. That marked the end of my career as a synthesis chemist."

The work for which Dr. Hoffmann is best known has to do with orbitals, the quantum-mechanical "wave functions" of electrons that surround the nuclei of atoms and which are responsible for all chemical interactions. A wave function of an electron is the probability of its being in any possible place. Wave functions can be positive or negative, they interact with each other, and they provide the glue that links atoms to form molecules.

In a molecule, the wave functions of the electrons of individual atoms interact with those of other atoms in patterns more complicated than those of a Rubik's Cube. In matching up orbitals to see if some desired pattern (that is, some hypothesized molecular structure) is possible, vast numbers of quantum-mechanical calculations are ordinarily needed. But by extending and expanding on a method devised in the 1930's by Erich Hückel, a German chemical theorist, Dr. Hoffmann found a way to make what he calls "very poor-quality but very useful" quantum calculations, which yield approximate predictions of how chemical interactions will proceed. These predictions sometimes turn out to be wrong, Dr. Hoffmann says, which is why the special insights of an experienced chemist are needed to distinguish probably correct results from implausible ones.

It is this kind of calculation that Dr. Hoffmann (and legions of other chemists) use today in attacking new reaction and synthesis problems. Sometimes Dr. Hoffmann and his students provide theoretical blueprints for creating interesting but useless molecules. At other times, the blueprints solve such highly practical puzzles as how the various components of electronic chips form chemical bonds with each other.

"It's difficult to explain our work to people who don't know quantum mechanics," he said. "But I sometimes use an analogy of a hostess seating people around a table who must all be matched up by sex or social position or something. You must have certain numbers of people and certain arrangements to make the seating come out right. It's somewhat the same with molecular orbitals."

This kind of research involves a great deal of fundamental physics, but Dr. Hoffmann vehemently denies that he is a physicist. "When I try to explain chemistry to outsiders," he said, "I have three main audiences: the person in the street, fellow academics in the humanities and physicists. All three audiences are equally ignorant of chemistry, but the most difficult audiences are the physicists, because they think they understand, but they don't."

He added: "Chemists don't have any holy grail, like the quest by particle physicists for the top quark or Higgs boson. Our goals are small but numerous, on the scale of the human condition itself."

A great shortcoming of theoretical physics, Dr. Hoffmann believes, is that physicists tend to be reductionists: they believe that everything in nature can be reduced to a few simple principles and particles. "Neither nature nor life works that way in reality," he said. "Complexity, not simplicity, is the essence of life. Take two molecules—say, dodecahedrane, a nice, simple, soccer-ball-shaped molecule, which has no use, and hemoglobin, an incredibly convoluted, complex molecule. For me, the hemoglobin is the more beautiful because of, and not despite, its complexity. Its complex form is essential for doing the complex things it has to do." Hemoglobin is the compound in red blood cells that chemically captures oxygen from the airways and transports it to the body's tissues.

Despite his frequent criticism of certain branches of physics, Dr. Hoffmann insists that he has no general bias against the field; his own daughter, Ingrid Zabel, received a Ph.D. in solid-state physics and is now doing research on Greenland's ice cap.

But Dr. Hoffmann chides some physicists for their concentration on the ultra-large—the entire universe—and the ultra-small—the fundamental particles making up atoms.

"Chemistry is an intermediate science between the extremes of size, more in consonance with the human scale of things," he said. "We chemists are down-to-earth people, who are not so troubled with the mysticism that sometimes creeps into physics. For example, I use quantum mechanics constantly, just as an engineer uses cement. But I don't agonize over its paradoxes the way some physicists do."

Dr. Hoffmann likes to take data from other experimental laboratories and show what it means in terms of quantum-mechanical structures. Based on laboratory measurements made by Dr. Douglas Rees and colleagues at the California Institute of Technology, Dr. Hoffmann and a graduate student, Haibin Deng, recently completed a theoretical analysis of the molecular interactions of molybdenum and iron in a natural enzyme called nitrogenase—the chemical that permits bacteria living in the roots of legumes to "fix" nitrogen from the air into compounds essential to plant life. "The molybdenum atoms at the edges of these iron clusters were the really weird ones to work out, but we've finally got a structural model for this enzyme, based on its molecular orbitals," he said.

In his quest for understanding, rather than for mere data, Dr. Hoffmann also believes that many things can be better explained by poetry or art than by equations.

"I take some heat from my chemist friends because I write poetry, which they consider just something to do when you're sulking," he said. "But they should take a look at the respective literatures of chemistry and poetry. The acceptance rate for scientific articles submitted to the best chemical journal in the world is about 60 percent. The acceptance rate for poems sent to even a mediocre poetry journal is about 5 percent. Furthermore, the poetry editors don't even give you a peer review and critique. They just turn you down flat."

He continued: "I guess I'd have to say that writing is difficult, but it has become nearly as important as research for me," he said. "Understanding is the important thing, and writing can help bridge the gap between science and the humanities. It can convey understanding."

—MALCOLM W. BROWNE
July 1993

LABORATORY OF GOOD AND EVIL

Roald Hoffmann's fascination with the complexities of chemistry has been sharpened by his appreciation of opposites—the complementarity of yin and yang in human behavior as well as in science.

In some of his writing, Dr. Hoffmann, a Jew who escaped Nazi Germany as a refugee, perceives a yin-yang admixture of good and evil in chemistry. For instance, he has written about Fritz Haber, a Nobel Prize winner who invented a process for capturing nitrogen from the air to make fertilizer needed to feed people—and explosives to kill them. Haber, a Jew who believed in the benevolent role of chemistry, also led the Kaiser's scientific team that developed the poison gases that poured death into the trenches of France during World War I.

Dr. Hoffmann is horrified by state-sanctioned cruelty wherever it occurs, and in the following poem he contrasts the changes that chemical substances undergo when compressed between two diamonds with the effects of torture on political prisoners in Argentina.

—M.B.

GIVING IN

At 1.4 million atmospheres
xenon, a gas, goes metallic.
Between squeezed single-bevel
diamond anvils, jagged bits
of graphite shot with a YAG

laser form spherules. No one
has seen liquid carbon. Try
to imagine that dense world
between ungiving diamonds
as the pressure mounts, and
the latticework of a salt
gives, nucleating at defects
a shift to a tighter order.
Try to see graphite boil. Try
to imagine a hand, in a press,
in a cellar in Buenos Aires,
a low-tech press, easily
turned with one hand, easily
cracking a finger in another
man's hand, the jagged bone
coming through, to be crushed
again. No. Go back up, up
like the deep diver with
a severed line, up, quickly,
to the orderly world of ruby
and hydrogen coloring near
metallization, but you hear
the scream in the cellar, don't
you, and the diver rises too fast.

—Roald Hoffmann

From *Chemistry Imagined,* by Roald Hoffmann and
Vivian Torrence, Smithsonian Institution Press,
Washington, 1993. This poem was first published
in *Paris Review* 33, No. 121 (1991):189.

Christiane Nüsslein-Volhard

"Lady of the Flies" Dives Into a New Pond

Agence France-Presse

File this one under "Starting your honeymoon with a root canal." Christiane Nüsslein-Volhard had recently learned the news most scientists do not dare fantasize about—that she had won the Nobel Prize—and she was hugely happy, as might be expected, and her laboratory mates at the Max Planck Institute for Developmental Biology were happy, and her peers in the developmental biology field were happy, not harvesting their sour grapes or grousing about so-and-so who had been left out, as often happens when these prizes are announced.

They knew that Dr. Nüsslein-Volhard, along with her co-winners, Eric Wieschaus of Princeton University and Edward Lewis of the California

Institute of Technology, richly deserved the honor for the work they had done in elucidating the earliest stages of embryonic growth by using the fruit fly, *Drosophila melanogaster.* People had been predicting Dr. Nüsslein-Volhard's Nobel for years, and nobody could doubt the importance of her contributions or the ferocity of her intelligence and drive.

Then came the figurative little trip to the dentist.

Dr. Nüsslein-Volhard had submitted for publication 24 papers about her laboratory's latest work, and now, in the standard process of peer review, she was getting feedback from other scientists. Their comments, however, were anything but the standard technical suggestions.

"I don't know whether this is courage, or foolishness," one reviewer wrote. "Why are you bothering?" wondered a second. And the most sputtering critique of the lot: "This is a terrible thing for science!"

"Can you imagine?" Dr. Nüsslein-Volhard said over lunch the other day in Boston, her voice rising in renewed anger at the memory. "What sort of comments are these? Do they think I'm stupid? Do they think I don't know what I'm doing?" Terrible for science, indeed. She complained to the journal editors about the patronizing and histrionic tone of her critics, and the papers were accepted for publication despite the carping, but the comments irritated her no end.

The sweetest honor, and a smack in the face, and all in an October's work. But perhaps this is what a scientist can expect when she leaves the cushiness of a project that has won her international renown and ventures into an entirely new arena. Several years ago, Dr. Nüsslein-Volhard, 53, traded in her title as reigning Lady of the Flies, Dame *Drosophila,* and instead took up the cause of a small striped animal called the zebra fish. She wanted to do the same sort of thing in the fish that she had done in the fruit fly: understand the basic patterns of development. She wanted to understand how a single fertilized egg could give rise to the complexity of something as exquisite as the zebra fish, an understanding that could eventually furnish answers to how human babies develop.

Yet while the work that led to her Nobel Prize in physiology or medicine had been built on the hoary wings of *Drosophila,* a laboratory staple since the beginning of the century, the zebra fish was still an experimental novelty, its ultimate value unproven in the field of genetics.

Dr. Nüsslein-Volhard is trying to push the field forward as fast as she can, and on a grand scale, but the skeptics are everywhere in evidence. When she gave a slide presentation of her latest work at a developmental biology conference in Boston, the first question from a member of the audience was, "You're an optimist, aren't you?" (Subtext: You'd better be.) Dr. Nüsslein-Volhard responded with humor. "One tries to be an optimist, of course," she said. "You must always say, 'This slide, isn't it gre-e-at?' "

Yet those who know her well say the dive from an established and over-populated field like *Drosophila* genetics into aqua incognito is in keeping with the restless, hungry intellect of the woman they affectionately call Janni (pronounced Yanni). "Moving into zebra fish was a bold move on Janni's part, and is consistent with her love of embryos, love of genetics, and love of new frontiers," said Dr. Judith Kimble, a developmental biologist at the University of Wisconsin in Madison. "I think it will pay off in the end, but it may take some time." One of Dr. Nüsslein-Volhard's greatest strengths, said Dr. Kimble, is that "she sets her sights on a truly big question."

Dr. Nüsslein-Volhard sees it slightly differently: "I must say, I'm easily bored."

Whatever the outcome of her fish work, Dr. Nüsslein-Volhard need not fear for her reputation. She has been famous for many years, long considered a shoo-in for a Nobel. Despite the arcane nature of her research, her home country of Germany has long recognized her as a national treasure. (One magazine profiled her in a roundup of "geniuses," though she was dismayed to see her picture near that of Adolf Hitler, the "genius of evil.")

Dr. Nüsslein-Volhard has a bracing, commanding presence that can either attract or intimidate people. At the Boston meeting, everybody wanted to claim acquaintance with her. It was "Janni this" and "Janni that," and she was repeatedly buttonholed by scientists seeking to collaborate with her. Some people regard her as haughty, but others say she is just the opposite, humorous and warm and generous with her time and insights.

"Janni is a charismatic and wonderful woman," Dr. Kimble said.

"I'm normal," Dr. Nüsslein-Volhard said. "I'm modest, but I'm also proud. I'm also a perfectionist and so I'm insecure. I get frustrated with myself easily."

Dr. Nüsslein-Volhard sometimes finds her fame to be a bit more trouble than it is worth. "The more famous you get, the more prizes you win, the more humble you have to be," she said wryly. "You have to put others at their ease. You have to assure them you're just like them, and that can be tedious sometimes." She also gets tired of having to travel here and there to collect prizes and then drag the prizes back home, but she admits that others may be unsympathetic to this lamentation. "People don't want to hear you complain about traveling to accept your awards," she said. "They don't want to hear about how I had to drag a box around with the Lasker award inside, and how that box felt like a rock." Now, she says, when an organization wants to give her a clever, unwieldy present, like the big 10-gallon Texas hat she was awarded recently, she tells them, just ship it.

Although some may say her latest work is risky and foolish, friends say Dr. Nüsslein-Volhard is anything but an unrealistic dreamer. "She has a very good sense of practical realities," said Dr. Wieschaus. "One of her favorite ways of dismissing other people is to say in German that they would not have survived the Stone Age."

The reasoning behind her decision to take up zebra fish was compelling. Fruit flies, as insects, are invertebrates, and invertebrates are structurally very different from vertebrates like fish, fowl and folk. The eyes of an invertebrate are different, its immune system is different, it has no central nervous system. Dr. Nüsslein-Volhard and other scientists contend that to truly understand vertebrate development, one must study vertebrates. Some are studying mice, but mice develop in the mother's womb, and thus a lot of the action happens out of sight. Other biologists study *Xenopus* frogs or chickens, but each of these animals has its technical limitations.

Zebra fish seemed like a perfect new "model organism" for prizing apart the details of vertebrate growth. The fish are small and so take up less space than a lot of vertebrates; they breed relatively rapidly, and the embryos develop outside the mother's body, as fish generally do. Moreover, the embryos are transparent, allowing a view of their development as it occurs.

For a long time after deciding to move into fish, Dr. Nüsslein-Volhard had to devote considerable energy to mechanical and janitorial matters like building the perfect aquarium, keeping the tanks clean, and figuring out what to feed the fish (fruit fly larvae, it turned out, are best).

Her institute, in Tübingen, Germany, was extremely supportive, building her a big new fish laboratory in which to stack her 7,000 tanks. Within the last year or so, she and her co-workers have been able to push into the next stage of the project, the real science—making mutants. That is what she has been reporting on lately—the 350 or so beautiful mutants that the laboratory has generated and begun to analyze.

The power of genetics lies in just this process: if you want to understand how genes work in development, you try to muck up those genes
and then see the results in the developing embryo. You use harsh chemicals or radiation to cause mutations in your subject animal, and then
observe how the offspring of the treated creature grow—or fail to grow.
The hallmark of a top-flight geneticist is the capacity to know a good
mutant when she or he sees one: to detect the slightest deviation from normal development and to recognize that the deviation is significant, that it
reveals something fundamental about the key steps in embryonic growth.

Dr. Nüsslein-Volhard and Dr. Wieschaus applied that logic to their collaborative work in fruit flies that eventually won them a Nobel, and Dr.
Nüsslein-Volhard is applying it again to zebra fish. She particularly likes
looking for mutants.

Dr. Daniel St Johnston of the Wellcome/CRC Institute in Cambridge,
England, a former postdoctoral fellow in Dr. Nüsslein-Volhard's laboratory who worked on fruit flies, said that she taught him to open his eyes
and see what was important. "Janni would be there late at night, looking
at mutants, teaching me how to look at *Drosophila* embryos," he said.
"Nothing would slip by her. She could spot the slightest deviation from
the norm and know whether or not it was significant." They would stare
together at a mutant through a microscope with two eyepieces, and she
would see if a single bristle on the embryo was, say, pointing in the wrong
direction.

The hurdles in the new work on zebra fish are manifold, the biggest
one being that right now, scientists have no way of isolating the mutant
fish genes for further analysis. Dr. Nüsslein-Volhard and her colleagues
have created 350 interesting mutants from 350 single genes to explain
their developmental defects, but have found no way to get those genes, to
clone them and to put them through their scientific paces, as is possible
with many genes in flies, mice and other organisms. Fish scientists are hin-
dered by a lack of even a crude gene map for the fish, a way of sorting out
one bit of piscine DNA from another. Dr. Nüsslein-Volhard concedes that
this is a problem, but she is confident it will be solved soon, as are the
many universities and institutions that are busily hiring young zebra fish
scientists. Not that Dr. Nüsslein-Volhard is particularly interested in solving the technical bottleneck herself; she finds the idea of gene mapping
and gene hunting frankly dull.

"I don't have a great tendency to clone genes," she said.

If Dr. Nüsslein-Volhard's skills are aesthetic as well as scientific, that is not surprising. She is the only scientist in a family of artists and architects. Her father was an architect, her mother painted and played a musical instrument, two of her four siblings are architects, and most of the children are amateur painters and musicians. Dr. Nüsslein-Volhard herself plays flute and sings lieder. "When my family gets together, we play music," she said. "We take our art seriously."

She decided at a young age that she wanted to be a scientist, which she says made it easier for her to pursue the profession in an age when women rarely did. "My family got used to it, my teachers got used to it," she said. "It was not a big deal. They thought, why not?" She glided relatively smoothly through graduate school at the University of Tübingen and post-doctoral fellowships in Basel, Switzerland, and Freiburg, Germany, only once encountering a male laboratory director who made her feel her sex was a drawback. She and Dr. Wieschaus collaborated on their extensive screens of fruit fly mutants in Heidelberg in the 1970's, and by 1980 the significance of their work was clear to scientific cognoscenti everywhere.

With the awarding of a Nobel Prize in science to Dr. Nüsslein-Volhard, the number of women so honored at last breaks into the double digits: 10. That figure is less than 3 percent of the total number of men who have been anointed laureates in the 94-year history of the award, yet it is high enough that most news reports of the Nobels did not bother mentioning Dr. Nüsslein-Volhard's chromosomal makeup, except in Germany, where she was celebrated as the first German woman to win.

But women must look elsewhere if they are seeking the cartoon role model who has checked off every box on the prescribed menu for female fulfillment: the career, the husband, the husband's career, the children. Dr. Nüsslein-Volhard was married briefly as a young woman, and kept the surname Volhard because it was already associated with her scientific reputation, but she has long been single and never had children. However, she is no scientific hermit, and she regularly holds dinner parties for her wide circle of friends. "She makes the best chocolate cake I have ever tasted," Dr. Kimble said.

The only monkish thing about her is her home. She lives in the millhouse of a 14th century monastery in a part of Tübingen so beautiful and becalming that Dr. Nüsslein-Volhard says she rarely needs vacations. When she wants to relax, she appropriately enough will go for a swim in the monks' original fishpond.

—Natalie Angier
December 1995

FASCINATING FISH

In the half-decade or so since Christiane Nüsslein-Volhard became the tenth woman to win a Nobel prize in science, the little zebra fish that she and other bold souls have championed as the research animal of the future has truly earned its stripes.

Dr. Nüsslein-Volhard won her renown and her Nobel through her studies of the embryonic growth of *Drosophila melanogaster,* the fruit fly, a laboratory staple since the turn of the 20th century. But as a woman with a formidable forebrain, and an impatience with the crossing of t's and the dotting of compound i's, she grew bored with her spineless flies, and decided that the best way to understand the development of vertebrate creatures like ourselves is to study a vertebrate animal. So in the early 1990's she dove into the relatively uncharted seas of zebra fish research.

Since then, hundreds of other researchers have also taken the plunge, with the result that zebra fish research is one of the most rapidly growing fields in biology. There is now a Zebrafish Genome Project, designed, like the more famous Human Genome Project, to map and analyze all the genes the animal possesses. There are dozens of zebra fish Web sites, zebra fish newsletters, zebra fish chat groups.

"The zebra fish is truly fascinating for an embryologist," said Dr. Nüsslein-Volhard in a recent interview. "The embryos are transparent, and easier to get and to watch than those of most other animals." As she sees it, the zebra fish is the ideal organism in which to study the development of the heart and the visual system, how an animal hears and balances itself, and even to make firmer links between genes and behavior.

One recent breakthrough in the field, she said, is the ability to add genes to an experimental fish that, when they are translated into proteins in the animal's cells, turn a bright fluorescent green and are blissfully easy to spot even as the fish swims in its tank and snacks on fly larvae. Hence, a researcher can insert a tagged gene into a fish embryo and then see in which organs—the liver, the muscles, the bones—the gene comes to life and goes about its task.

Winning the Nobel Prize was a mixed blessing for Dr. Nüsslein-Volhard. "Overall, I really loved it," she said. "But it certainly did change the attitude of many people towards me. I didn't change much, I think, but I am provoking a lot of people to react against me. They seem to be motivated by envy and jealousy to a surprising degree."

On the other hand, she is pleased that she has been an inspiration for a number of her young relatives. "One of my nephews is an assistant professor of chemistry at the Scripps Clinic in San Diego," she said. "He participated in the celebrations in Stockholm, and I'm sure they made a deep impression on him that won't be easy to get away from!"

Dr. Nüsslein-Volhard also acknowledges that she is a role model for young women generally, although she says that "progress is slow indeed" for women in science. After all, she might have finally pushed the number of female science laureates into the double digits in 1995, but laureate No. 11 has yet to show her face.

—N.A.

Stephen W. Hawking

Sailing a Wheelchair to the End of Time

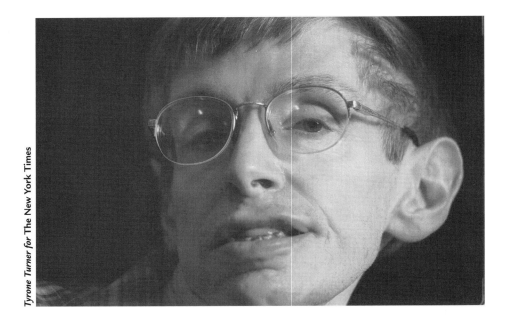

Tyrone Turner for The New York Times

Out of a crumpled, voiceless body in a wheelchair, the mind of Stephen W. Hawking soars and summons expression by pressing a finger and thumb to a small control box in his lap. Slowly, one word or letter at a time, sentences appear on the lower half of a compact computer screen attached to the chair. He is forming an answer to a question about the consuming quest of some of the world's most brilliant scientists, himself very much included, to find a "theory of everything" to explain all phenomena in space-time, especially the first split second of cosmic creation in the Big Bang.

Such a theory would unify the two pillars of 20th century physics, Albert Einstein's general theory of relativity and the quantum theory. Relativity is a theory of gravity dealing with the macroworlds of stars and planets; quantum theory describes the strange microscopic properties of elementary matter. To understand the early universe, when everything was unimaginably small and densely packed, one needs a combined quantum theory of gravity, the long-dreamed-of theory of everything.

Lately, Dr. Hawking has been saying that success may be in sight. Why the optimism? Will quantum gravity be a Hawking theory?

His head lolling, Dr. Hawking appears half-asleep in concentration over the answer. His wide mouth grimaces at the physical and mental exertions. Then he grins at some jocular turn in his thoughts, his blue eyes brightening behind gold-rimmed glasses. Finally, he taps a command to the computer-driven voice synthesizer. The man many consider the foremost cosmologist of the day, an intellectual superstar of the first magnitude, is about to "speak."

The disembodied voice is cyber-surreal. "My wife says I'm a born optimist," he jokes, "so maybe you shouldn't believe me."

But people seem to want to believe that Dr. Hawking, if anyone, has answers about the nature of space and time, the beginning of the universe and its fate. In a recent visit to the United States, he drew large crowds at the White House and for lectures at the California Institute of Technology in Pasadena. Whether President Bill Clinton and his guests, eminent university scientists or an interviewer, the audiences listen expectantly.

There is a Delphic authority to the unworldly synthetic voice, and a promise of revealing insights conveyed in his even more abstract language and mathematical constructs of cosmology. It must have been something like this when the ancients gathered to partake of the lofty wisdom of their oracles.

This oracle usually resides at Cambridge University in England, where he occupies the professorial chair once held by Sir Isaac Newton. From his wheelchair, Dr. Hawking remarks with typical humor, "They say it's Newton's chair, but obviously it's been changed."

Part of Dr. Hawking's celebrity no doubt derives from his pluck in the face of physical adversity. When he was a graduate student at Cambridge in 1963, doctors predicted an early death from the incurable neuromuscular disorder commonly known in the United States as Lou Gehrig's disease.

He has since lost the use of his legs and arms. He cannot feed himself, and an operation in 1985 to assist his breathing cost him what little remaining use he had of his vocal cords. Only the flicker of energy in his hands allows him to operate the motorized wheelchair and to communicate ideas and lectures with the computerized voice synthesizer. He prepares his lectures in advance on the computer and then, from his wheelchair on stage, controls the delivery by the voice synthesizer, pausing frequently to let an idea sink in or wait for a laugh.

Dr. Hawking has told colleagues that in a way his illness had been a blessing; it made him focus more resolutely on what he could do with his life. Even so, his affliction restricts his research methods.

"I avoid problems with a lot of equations or translate them into problems of geometry," he said recently. "I can then picture them in my mind."

Now 56, Dr. Hawking has not only survived but established an exalted scientific reputation. He was inducted into the Royal Society at an unusually young age and has since been knighted. Sometimes he is spoken of in the same breath with Newton and Einstein. Asked about this at the White House, he called such comparisons "media hype," which brought uproarious applause from a political audience more familiar with this popular phenomenon than with quantum gravity.

"I think I fit the popular stereotype of a mad scientist or a disabled genius or, I should say, a physically challenged genius, to be politically correct," Dr. Hawking continued. "I don't feel I am a genius like Newton or Einstein."

But his early scientific celebrity rested securely on his explorations of the implications of relativity theory that went beyond Einstein's own imagination. This led to breakthrough concepts of the nature and behavior of black holes, those voracious gravitational sinkholes in space predicted by relativity but never really accepted by Einstein. Recent observations, indirect but increasingly persuasive, point to the existence of many black holes with masses the equivalent of millions of Suns.

Working with Dr. Roger Penrose, now at Oxford University, Dr. Hawking considered the problem of what happens when enormous amounts of stellar matter collapse into a black hole, as predicted by relativity. They concluded that the collapse must continue into an increasingly smaller and denser point called a singularity. At such a point, where the density would be infinite, all the basic laws of physics would have broken down.

Dr. Hawking took this insight a step further in the late 1960's. As he saw it, the expansion of the universe must be like the time reverse of the collapse of matter creating a black hole; running the process backwards should lead to a beginning. In that case, if Einstein's math is correct, the primeval atom out of which the universe supposedly emerged in the Big Bang must have been the mother of all singularities.

The idea, hailed as a critical underpinning for the Big Bang, is now generally accepted.

Another of his major insights was that black holes are not entirely black. These places were thought of as black because their gravitational attraction is so powerful that it consumes surrounding matter and does not permit the escape of any light. But while trying to disprove an idea of Dr. Jacob Bekenstein, then at Princeton University, Dr. Hawking wound up confirming and improving on it by describing how through quantum mechanics black holes emit fleeting particles in the region of their outer boundary.

Through Hawking radiation, as this is called, a black hole leaks some energy and will presumably diminish in size and probably explode in some distant future.

The faint thermal radiation should make some black holes theoretically detectable, though observations so far draw a blank. Nevertheless, Hawking radiation became an established property of the cosmos, once initial skepticism was overcome.

At one of his Caltech lectures, Dr. Hawking recalled a visit to France soon after he made the discovery. "My seminar fell rather flat, because almost no one in Paris believed in black holes at the time," he said. "Maybe it was the name, which the French considered obscene, and refused to use."

Pondering what it is that sets Dr. Hawking apart from most other theoretical physicists, Dr. Kip S. Thorne, a longtime friend on the Caltech faculty and no mean theorist himself, said in an interview: "Stephen can see much farther and much more quickly what nature is likely to be doing than most of the rest of us poor mortals. Very few have his level of understanding and insight, or his ability to ask the right questions that trigger others to work on problems in ways they might never have thought of."

Dr. Hawking's provocative work, for example, has inspired many new interpretations of Einstein's theory and the search for the so-called theory of everything. The discovery of Hawking radiation, Dr. Thorne said, was "a tremendous surprise and really the first step toward a marriage between relativity and quantum theory, toward a theory of quantum gravity."

Another Caltech physicist, Dr. John P. Preskill, says Dr. Hawking is "not a linear thinker, but is very good at taking intuitive leaps and then filling in the details" or letting others do it. From time to time, the details or new observations do not support one of his theorems. "He's not always right," Dr. Preskill said, noting that his recent lectures reflected a reevaluation of some of his ideas in response to new findings.

This is a problem for all theoretical cosmologists these days, not just Dr. Hawking. With far-seeing telescopes on the ground and in space, every few weeks astronomers present cosmologists with a new set of data that rein in one theory or another.

Increasing evidence indicates that the average mass density of the universe falls well below levels needed to confirm cosmologists' favorite models. Earlier this year, measurements of distant exploding stars clocked the expansion of the universe and found that it seemed to be speeding up, contrary to most assumptions.

Something besides the gravity of matter, which should be slowing the expansion rate over time, must be acting to counteract gravity. Could it be energy from the vacuum of space that creates a repulsive force, called the cosmological constant?

Einstein first proposed such a force, or negative pressure produced by vacuum energy, to correct what he thought was a flaw in his relativity theory. Later, he recanted when observations showing an expanding universe appeared to render the theorized force unnecessary. More recent findings that suggest a low-density universe, however, have brought the cosmological constant back in fashion.

Only now does Dr. Hawking reluctantly agree that observational evidence seems to favor Einstein's old idea. Giving in with a characteristic dash of whimsy, he said, "Negative pressure is just tension, which is a common condition in the modern world."

Dr. Hawking seems to engage in some nimble geometry to preserve another of his concepts, a "no-boundary universe," which he developed with Dr. James Hartle of the University of California at Santa Barbara. They pictured a finite, closed universe in the shape of a sphere, only in four dimensions. It starts with a Big Bang, expands to a maximum point like the spherical Earth's Equator and then contracts toward an eventual collapse in what is sometimes called the Big Crunch. Like Earth's surface, this model has no edges and would seem to require a closed universe.

Talking to the Caltech faculty, Dr. Hawking defended the concept, even though current findings indicate an "open" universe with a concave rather than convex geometry. He and Dr. Neil Turok of Cambridge, he said, had conceived of another way of looking at the no-boundary universe as either spherical and thus closed and finite or open and infinite, in which case its shape would resemble an open horn.

How could the universe be in two distinctly different shapes?

Physics has dealt with such problems before, notably the debate over whether light was a wave or a particle. Eventually, physicists determined that light is both wave and particle. It all depends on one's point of view.

So Dr. Hawking, in rethinking the no-boundary universe, has conceived of it as having extra dimensions in space-time, including an "imaginary time" running at right angles to real time. Such a universe, Dr. Hawking and Dr. Turok theorized, would have more possibilities. It could be shaped like a cone. Sliced horizontally, it would look like a circle, or a closed universe. Sliced vertically, it would be a parabola, or an open universe.

By seeking to preserve the no-boundary universe, Dr. Hawking was clinging to a concept that did not require an appeal to something outside the universe, beyond the laws of physics, to determine how the universe began.

His revised thinking even includes an idea of how the universe came into being in the form of an extremely small particle, a kind of genesis particle of space and time. He calls it a pea instanton, because it would presumably be a slightly irregular sphere like a pea.

Scientists who heard the talk said that an instanton should be thought of as not so much a thing as an event. It was comforting to hear Dr. Preskill, a quantum physicist, remark that probably few in the audience, not even the professors, had comprehended exactly what Dr. Hawking was talking about.

"Stephen may be reaching here, going out on a limb," Dr. Preskill said. It would not be the first time. As colleagues said, Dr. Hawking often goes to the edge with his ideas, hoping to stimulate others and thus get feedback for fine-tuning his own thinking.

If the universe did begin as a kind of four-dimensional wrinkled pea, American and European spacecraft planned for the first decade of the next century may find confirming evidence in the most detailed examination yet of potentially revealing temperature fluctuations in the cosmic microwave background, the faint echo from the Big Bang.

In a revealing aside in one lecture, Dr. Hawking seemed close to throwing in the towel on one of his black hole theorems. He had once proposed that a multitude of tiny black holes, each with a mass no greater than Mount Everest, just might have been left by forces immediately after the Big Bang. Their quantum temperatures would be swamped by the cosmic background radiation, but by now, if the theorem is correct, many of them should be exploding, issuing possibly detectable gamma rays.

"There don't seem to be many of them around," Dr. Hawking said of the mini–black holes. "That is a pity. If one were discovered, I would get a Nobel Prize."

Nobel Prizes do not usually go to discoveries in observational astronomy, though there have been exceptions in the cases of pulsars and the cosmic microwave background. In theoretical astronomy, Dr. Preskill said, the prize is usually bestowed only "in connection with some fairly direct and dramatic confirmation of a particular prediction."

Dr. Hawking's growing preoccupation these days is the search for a way to understand more clearly space and time, especially the initial conditions, through a union between relativity and quantum mechanics. Einstein made major contributions to quantum theory, though he never really liked it. But he devoted his last decades to thinking about a unifying theory, to no avail. Dr. Hawking jokes that that is what aging physicists like himself do: look for a theory of quantum gravity.

For once Dr. Hawking may not be in the vanguard of thinking. "He's very much a player, but isn't one of the leaders," said Dr. Craig J. Hogan, an astrophysicist at the University of Washington in Seattle.

The most promising area of research is string theory, a highly abstract concept floated in the 1960's and now recognized as a possible basis for a theory of everything. Strings are thought of as tiny vibrating loops that are not matter or energy, but rather are the elements out of which those things were made.

In the last three years, physicists have come to realize that various versions of string theory are linked to one another in subtle ways. Moreover, both quantum mechanics and general relativity are thought to be special cases of this theory.

"It smells and feels like the answer to quantum gravity is coming, whether on a time scale of 2 years or 20 years is uncertain," Dr. Thorne said. "We're certainly not talking about 100 years."

Though Dr. Hawking stimulated the pursuit with his research on singularities and black-hole behavior, Dr. Preskill said, "he's converging more and more with people who are trying to understand quantum gravity by string theory and particle theory."

Backstage after a lecture to the Caltech faculty, Dr. Hawking is exhausted. His nurse looks on with some concern. She wipes his mouth and whispers something. He agrees to go ahead with the interview, taking only a few impromptu questions. Later, he would provide computer-written answers to other questions that, as requested, had been submitted in advance.

To the question about the prospects for finding a theory of everything, Dr. Hawking says that his optimism for success in the next 20 years stems from "tremendous progress" in recent years toward understanding interactions between particle physics and gravity, quantum theory and relativity. He is further encouraged, he says, by the knowledge that there is an irreducibly smallest interval of time and space, known as the Planck length, which would apply in the very early stages of the universe when everything was incredibly small.

"It seems that there is an ultimate length scale that we can't go below, so there ought to be an ultimate theory," he concludes. "If we find it, it will be through the combined efforts of large numbers of people, not just one individual."

Still, people gather to hear the thoughts from this soaring mind in a wasted body bound to a wheelchair. A Caltech official, seeing the overflow crowds for the public lecture, was reminded of the turnouts for Einstein on his visits to the campus years ago.

Perhaps the crowds come out of a sense that Stephen Hawking is a living metaphor for that sublime expression of human audacity: the belief that the universe is comprehensible by a species with its own brief history of time and physically confined to one small world and its immediate environs, but with a vision that can encompass some 15 billion years of space-time.

—JOHN NOBLE WILFORD
March 1998

INCHING TOWARD A THEORY OF EVERYTHING

The pursuit of a theory of everything quickened in the months after this interview with Dr. Stephen Hawking. Developments in string theory seem—for now, at least—to justify Dr. Hawking's optimism for success in the next 20 years in recognizing a quantum theory of gravity, the so-called theory of everything that unites the forces operating on the microscopic scale of quantum mechanics with the large-scale force of gravity.

Even if Dr. Hawking is not a leader in string theory, those who are credit him as one of the sources of an insight that has reinvigorated the theory, which postulates that the universe is made from tiny, vibrating stringlike particles and multidimensional membranes. For several decades, the theory had existed as little more than sets of equations in search of fundamental principles and some compelling predictions that could be tested by experiment.

Now, some physicists think that they may find in black holes the demystifying Rosetta stone of their dreams. These are not the massive gravitational sinkholes that appear to exist at the cores of many galaxies, but tiny hypothetical objects resembling elementary particles. In the physics of these objects, the same phenomena may be written in the languages of both quantum mechanics and general relativity.

"These are the two great achievements of 20th century physics," Dr. Andrew Strominger of Harvard University said in an interview with the journal *Science* (July 23, 1999), "and for the first time we're seeing, at least in some cases, that they are really two sides of the same coin."

About 25 years ago, as the *Science* article pointed out, Dr. Hawking and Dr. Jacob Bekenstein, now of Hebrew University in Jerusalem, examined the dynamics of black holes and recognized behavior implying that black holes had a microscopic description. They described how through quantum mechanics black holes emit fleeting particles at their outer boundaries. It was just one of many examples of Dr. Hawking's provocative ideas about black holes. In this case, the article concluded, scientists may have found "a potential bridge between the macro- and microworld."

No one is saying that this and other recent advances in string theory have brought cosmology to the threshold of a unifying theory of everything. But these are encouraging steps.

They also seem to bear out something else Dr. Hawking said in the interview about a theory of everything: "If we find it, it will be through the combined efforts of large numbers of people, not just one individual."

For the practice of scientific creativity is changing, largely through the electronic publication of research papers. Within hours after authors complete a paper on their new idea, other scientists around the world can read it on the Internet and weigh in with their own contributions.

"As a result, no one theorist or even a collaboration does definitive work," the *Science* article noted. "Instead, the field progresses like a jazz performance. A few theorists develop a theme, which others quickly take up and elaborate. By the time it's fully developed, a few dozen physicists, working anywhere from

Princeton to Bombay to the beaches of Santa Barbara, may have played important parts."

Add to that the crippled genius in a wheelchair at Cambridge University. An unforgettable presence, Dr. Hawking still manages to travel regularly to international conferences and share the prodigious output of his mind with other cosmologists trying to understand how the universe works.

—J.N.W.

Aaron T. Beck

Pragmatist Embodies His No-Nonsense Therapy

The session, Dr. Aaron T. Beck recalls, began like many others. The woman lay on the couch, describing her sexual encounters with men, while Dr. Beck, at the time a recent graduate of the Philadelphia Psycho-analytic Institute, sat behind her, scribbling in his notebook.

"How does talking about this make you feel?" he asked her.

"I feel anxious," she replied.

Trained to probe the hidden conflicts underlying psychological symptoms, Dr. Beck responded with an interpretation.

"You are anxious because you are having to confront some of your sexual desires," he told her. "And you are anxious because you expect me to be disapproving of these desires."

"Actually, Dr. Beck," his patient replied, "I'm afraid that I'm boring you."

Arms crossed on his chest, red bow tie resplendent, pale blue eyes keen beneath a shock of white hair, the founder of the fastest growing, most extensively studied form of psychotherapy in America is telling this story to explain how he eventually came to leave Freud behind.

Sitting in his office at the Beck Institute for Cognitive Therapy and Research in Bala Cynwyd, outside Philadelphia, he offers a favorite maxim: "There is more to the surface than meets the eye."

The key to many psychological difficulties, Dr. Beck has found in 40 years of research and clinical work, lies not deep in the unconscious, but in "thinking problems" that are much closer to conscious awareness.

In the woman's case, for example, it turned out that she engaged in an endless self-deprecating monologue, an inner voice constantly berating her, telling her that she was unattractive, uninteresting and worthless.

And these "automatic thoughts," as Dr. Beck calls them, led her to behave in self-defeating ways, like acting promiscuously because she did not think she had much else to offer, or engaging in histrionics in an effort to seem more interesting.

Cognitive therapy, developed by Dr. Beck after he abandoned psychoanalysis, is intended to help patients correct such distortions in thinking, often in a dozen sessions or fewer.

Dr. Beck calls the method "simple and prosaic," with no dredging up of lost childhood memories, no minute examination of parental misdeeds, no search for hidden meanings.

"It has to do with common-sense problems that people have," he said.

Patients in cognitive therapy are encouraged to test their perceptions of themselves and others, as if they were scientists testing hypotheses. They receive homework assignments from their therapists. They learn to identify their "inaccurate" beliefs and to set goals for changing their behavior.

It is an appealing package. And in an age when managed care closely monitors the consulting room, and most psychiatrists view drugs—not talking—as the treatment of choice for their patients, Dr. Beck's approach has been able to provide hard data in support of psychotherapy's power.

Cognitive therapy's basic precepts are easily summarized in training manuals, and its simplicity makes it an ideal research tool. And dozens of studies have shown it to be effective in treating depression, panic attacks, addictions, eating disorders and other psychiatric conditions. Researchers are also studying the therapy's ability to treat personality disorders and, in combination with drugs, psychotic illnesses like schizophrenia.

Therapists from around the world travel to the Beck Institute for training. And mental health organizations like the National Mental Health Association recommend cognitive therapy to patients as one of the few forms of psychotherapy studied in large-scale clinical trials.

Yet every theory of the human mind in general springs from a human mind in particular. Freud, caught in his own Oedipal struggles, saw the unconscious as roiling with sexual and aggressive impulses. Fritz Perls, possessed of a biting wit and fond of confrontation, invited his patients to take the "hot seat." Carl Rogers, a former seminarian and by all accounts an empathic soul, argued that psychotherapy should be "client-centered."

And in its way, cognitive therapy—practical, cerebral and to the point—is also a fair reflection of the man who conceived it.

He is 78 now, an emeritus professor of psychiatry at the University of Pennsylvania, four times a father, eight times a grandfather. Yet even as a younger man, his former students say, Dr. Beck, with his white hair and the bow tie he carefully put on each morning, projected a grandfatherly air, offering a nurturing presence, a passion for collecting data, a conviction that evidence always trumps opinion.

Others in his position might cultivate the flamboyance Americans seem to expect of their therapy gurus. But Dr. Beck has more in common with Marcus Welby than Dr. Laura Schlessinger or John Bradshaw—his currency is ideas, not personal charisma. Soft-spoken and unexcitable, he wears a hat, chats amiably with strangers in elevators and uses words like "gosh" and "gal."

Asked to describe himself, Dr. Beck ticks off "kind, intelligent, creative, flexible."

"I don't need to be right," he says, "but I don't like to be wrong."

Dr. Jeffrey Young, a former student, now the director of the Cognitive Therapy Center of New York, recalls a debate with his professor over whether those who came to them seeking help should be referred to as "patients" or "clients." Dr. Beck had a simple solution: Ask people what term they prefer.

"I think I am ultimately a pragmatist," Dr. Beck says, "and if it doesn't work, I don't do it."

He encourages a similar philosophy in his patients, hoping they will eventually choose to let go of the self-defeating attitudes that tie their lives in knots. "It's a testable assumption," Dr. Beck tells a 30-year-old woman who believes, she told him, that "if I don't punish myself, God will be mad."

"You could see if you stopped punishing yourself and nothing happened," he suggests.

With patients convinced that they must always be perfect, that their bosses hate them, that their spouses are insensitive to their needs, he will question, gently, "Would you agree that it is against your best interests to have this belief?" He will ask: "What are the disadvantages to thinking this way?" He will wonder out loud: "Do you think it is possible to ignore these thoughts?"

It is a faith in the rational mind he has carried since childhood, growing up in a middle-class neighborhood of Providence, Rhode Island, the third son of Russian Jewish immigrants, his father a printer with strong socialist beliefs who wrote poetry in his later years, his mother a forceful woman of unpredictable moods who had already lost two children.

He was a Boy Scout, an active child who, despite his mother's overprotectiveness, played football and basketball until at 8, he developed a dangerous staph infection after surgery for a broken arm, a complication that kept him in the hospital for more than a month.

He remembers the surgeon saying "He's not under yet," remembers a terrible dream of a series of alligators, each biting the tail of the next, the last alligator biting his arm.

He remembers his mother saying: "He will not die. He will not die." The boy himself never questioned that he would recover. But the surgery, Dr. Beck believes in retrospect, was a defining moment in his life, restricting his activities and forcing him to find quieter forms of entertainment, like reading.

The hospitalization defined his life in other ways, too. He developed a phobia of blood and injury: a hospital scene in a movie was enough to send his blood pressure plunging. If he smelled ether, he became anxious and began to faint.

He conquered his fears methodically, allowing logic to gradually triumph over irrationality. "I learned not to be concerned about the faint feeling, but just to keep active," he says.

With such a straightforward attitude toward his own psychology, Dr. Beck was probably never meant to become a psychoanalyst; even now, his interest in how his childhood experiences shaped him seems minimal. Freudian theory was ascendant in psychiatry departments across the country when he was a resident at the Cushing Veterans Administration Hospital in Framingham, Massachusetts. And like many of his peers, he pursued analytic training, graduating from the Philadelphia institute in 1958.

Still, he had some doubts. The lack of precision annoyed him: though every analyst agreed that in neurosis there were "deep factors at work," no one, Dr. Beck discovered, could agree on exactly what those factors were.

He found work with patients exhausting, because the goals seemed so unclear. "The idea was that if you sat back and listened and said 'Ah-hah,' somehow secrets would come out," Dr. Beck remembers. "And you would get exhausted just from the helplessness of it."

He completed his training and began taking patients in for analysis. But without any fanfare, he began to adjust the way he interacted with them. The woman who worried about boring him, for example, he asked to sit up and face him, so that she could see his facial expressions and gauge his interest in what she was saying. He began to ask more questions, and to listen to the answers in a different way.

At the same time, at Penn, where he joined the faculty in psychiatry in 1954, Dr. Beck was trying to find empirical evidence for Freudian precepts—and failing. With a colleague, he designed an experiment to test the link between depression and masochism, a basic psychoanalytic notion. But the researchers found no evidence that the depressed patients in the study somehow needed to suffer. Instead, Dr. Beck said, they simply showed low self-esteem, devoid of hidden motives. "They saw themselves as losers because that's the way they saw themselves," he said.

The cognitive approach to therapy that Dr. Beck ultimately developed—influenced, he says, by thinkers like Karen Horney, George Kelly and Albert Ellis, whose rational emotive therapy struck similar themes—was a major departure from the psychoanalytic fold. And it was not received warmly. Many analysts dismissed it as superficial; some suggested that perhaps Dr. Beck himself "had not been well analyzed."

There have been other critics, as well. Psychologists trained in classical behaviorism have opposed cognitive therapy's focus on "thoughts," which they said could not be measured objectively. Biological psychiatrists, like Dr. Donald Klein, director of research at New York State Psychiatric Institute, have argued that studies testing the therapy's effectiveness lacked adequate scientific controls. If the technique works, Dr. Klein has suggested, it is because it acts as a morale booster for "demoralized" patients, not because it offers specific treatment.

Dr. Beck, for his part, has responded to each critique with a new raft of experimental data.

"He is an unusual person," said Dr. John Rush, professor of psychiatry at the University of Texas Southwestern Medical Center and a former student. "He is willing to test his own beliefs, just like he asks patients to test theirs."

Yet in the early years it often was lonely work, and it was his wife, Phyllis, now a Superior Court judge in Philadelphia, who buoyed him.

"She was my reality tester," he said. "She went along with the newer ideas I had, and that gave me the idea that I wasn't in left field."

Many decades later, she remains his closest confidante. But it is his daughter, Dr. Judith Beck, a psychologist who is director of the Beck Institute, who participates most closely in his work.

Scene: a suburban delicatessen, a corned beef sandwich, his daughter sitting next to him; a comfortable setting for Dr. Beck who, his colleagues and former students say, is in fact very shy.

"Do you remember that dream I had when I was going off to graduate school?" she asks him. "That I was up on the Empire State Building and I felt in danger of falling off."

"I do," he says. "And do you remember what I told you it might be about? That the higher you aspire, the greater you're going to fall?"

"It hit me as absolutely that was what it meant," she replies.

As institute director, she has come to know her father in a different way, to admire him as a thinker and a therapist, to work with him as a colleague. When she was a child, she says, he was always working; age has made him more tolerant, less driven, has turned him more toward family.

It has not slowed him down. He receives 10,000 e-mail messages a year, divides his time between Penn and the institute, is expanding his research into new areas. He plays tennis regularly, despite a recent hip replacement. His newest book, *Prisoners of Hate: The Cognitive Basis of Anger, Hostility and Violence* (HarperCollins, 1999), appeared last fall.

Retiring, he says, has never entered his mind.

"I think he has done a lot of cognitive therapy on himself," his daughter says.

—ERICA GOODE
January 2000

HELPING THE TOUGHER CASES

Cognitive therapy was developed 40 years ago to treat people suffering from depression.

But in the age of Prozac and other newer antidepressants, said Dr. Judith Beck, director of the Beck Institute for Cognitive Therapy and Research, "we don't see them in our offices anymore."

The patients who do seek cognitive therapy these days tend to have more long-standing—and more complicated—problems. And in response, the therapy is being modified and adapted to meet their needs.

In treating borderline personality disorder, for example, a cognitive therapist may ask patients more about their childhoods, hoping to find the "early conditioning experiences" that helped nourish their distorted beliefs about themselves and others.

And where someone with simple depression is likely to improve in 8 to 10 sessions with a therapist, said Dr. Aaron Beck, the founder of cognitive therapy and Dr. Judith Beck's father, patients whose problems are more global may remain in therapy for several months, a year, or longer.

One goal in such cases, said the senior Dr. Beck, is "to try to teach them self-control, how to control their impulses."

The relationship with the therapist also becomes more important than in shorter-term therapies. For example, Dr. Beck said that a 30-year-old woman who sought help at the Beck Institute's clinic initially saw him both as an authority figure who would try to control her, and as a helper who had her best interests at heart.

His strategy in such cases, he said, is to talk to patients about their beliefs, and invite them to test out their perceptions, to see if they mesh with reality. If the patient believed he was attempting to control her, for instance, he might ask: "How would you expect me to behave if that were case?" and "What is the evidence in favor of this? What is the evidence against it?"

It is a method that Dr. Beck believes can be helpful even with patients with severe psychotic disorders, like schizophrenia.

In the United States, treatment for schizophrenia is generally limited to the use of anti-psychotic drugs, perhaps with addition of supportive counseling to help patients and family members cope. But Dr. Beck and other researchers are finding that when added to drug treatment, cognitive therapy can help psychotic patients gain more control over their hallucinations and delusions.

Seven studies in England, Canada and Italy, Dr. Beck noted, have shown cognitive therapy to be effective both for chronically ill patients who do not respond to drugs, and for patients in the throes of acute psychotic symptoms.

In a review of the research, not yet published, Dr. Beck and Dr. Neil A. Rector, of the University of Toronto, concluded that patients with schizophrenia who improved through cognitive therapy "continue to experience fewer distressing symptoms, have lower relapse rates, spend less time in hospital, and appear to have greater skills to negotiate setbacks than patients receiving routine care alone."

Cognitive therapists use many of the same techniques to treat psychotic patients as they do in treating patients with less severe illnesses. But therapy sessions tend to be shorter and the treatment extended over a longer period of time, homework tasks are more focused and goals are more flexible, Dr. Beck and Dr. Rector noted.

The therapy, they pointed out, is not intended to "cure" delusions or hallucinations, but to reduce the distress they cause, for example, by challenging patients' beliefs that the voices they hear are omnipotent and cannot be disobeyed.

"The goal is to render the experience less threatening by altering the meanings associated with voices, rather than diminishing the hallucinatory behavior itself," the researchers wrote.

Cognitive therapy may work in schizophrenia, Dr. Beck speculated, because it helps patients gain access to their abilities to think logically and to organize their mental processes.

In a session with a man convinced that his neighbors were tormenting him by banging on the walls, for example, Dr. Beck suggested that he think of himself as a valuable person, above paying attention to such distractions.

The patient was, for a moment, speechless. "No one has ever called me valuable before," he said.

—E.G.

Edward O. Wilson

From Ants to Ethics: A Biologist Dreams of Unity of Knowledge

Rick Friedman *for* The New York Times

Laid out on a desk in Harvard's Museum of Comparative Zoology is a neat array of white-sided boxes, each with pinned specimens of ants belonging to the genus *Pheidole*. Dr. Edward O. Wilson is writing a monograph on the genus, which is the world's largest. There are 326 new species, many of which he discovered himself, and each is in need of a precise description and a scientific name. The work is pure taxonomy, "scientific knitting," he calls it. The ants, little black specks mounted on white triangles of cardboard, seem perfectly indistinguishable. But under the microscope each species looks as different as a bear does from a tiger, Dr. Wilson assures a visitor.

The "knitting" is no doubt mental relaxation from the soaring works of synthesis that are Dr. Wilson's other passion. Or maybe both are the products, on different scales, of a mind that loves to classify and discover patterns in the world's unruly substance.

In an alternative fate, Dr. Wilson might have been an obscure expert on the ants of Alabama, his home state. But at each stage of his career he has looked outward, trying to see how the scholarly patch he had cultivated might fit into some larger scheme of things. And because so few scholars dare to explore beyond the boundaries of their own narrow fields, Dr. Wilson has produced an original work of synthesis time after time.

His first foray outside the ant microcosm was *The Insect Societies*, published in 1971, which surveyed the evolution of sociality among wasps, ants, bees and termites. It seemed only logical to extend the synthesis to vertebrates, humans included, though no one else had attempted so obvious a task. *Sociobiology* appeared in 1975, immediately falling under political attack from left-wing colleagues who objected to the idea that certain human behaviors might have genetic determinants.

"He was taken to task in an unfair way," says Dr. Thomas Eisner, a Cornell University biologist and longtime friend, "but his ideas are not arrived at lightly." The ideas of sociobiology were many years in the making, Dr. Eisner says.

Dr. Wilson was blindsided and bruised by the attack but undeterred. He expanded his ideas about the mind's genetic history in *On Human Nature,* which appeared in 1978. Then he turned to other matters, like writing a comprehensive survey of ants with his colleague Bert Holldobler and becoming a champion for the world's disappearing rain forests.

Twenty years later, Dr. Wilson has returned to the theme of genes and human nature. His new work of synthesis is called, a little forbiddingly, *Consilience,* a word coined by the 19th-century philosopher of science William Whewell to mean the melding of inferences drawn from separate subjects. Dr. Wilson has resurrected it as the slogan for a program of unrivaled ambition: to unify all the major branches of knowledge—sociology, economics, the arts and religion—under the banner of science and in particular of the biology that has shaped the human mind.

The kind of unification he proposes is the outright intellectual annexation that occurs when one field of knowledge becomes explainable in terms of a more fundamental discipline, as when thermodynamics (heat processes) was explained in terms of statistical mechanics (equations describing the movement of molecules). These putsches, which always constitute major advances yet are usually not much welcomed by scholars in the discipline being taken over, are termed "reductions" by philosophers of science.

How can economics, not to mention religion and aesthetics, be reduced to molecular biology? Dr. Wilson believes this will come about as biologists work out the behavioral rules that evolution has built into the brain. These rules, the sum of which is human nature, define the framework in which economic, religious and aesthetic decisions are made and so should, he says, be made fundamental to those branches of study.

Dr. Wilson is well aware that people do not always warm to the idea of having their higher cognitive functions explained in terms of genetic programming. But he can see the revolution coming, the vanguard being led by the new approach to the brain called evolutionary psychology (another name for sociobiology, in his view).

Karl Marx, Dr. Wilson once joked when talking about ants, was correct: he just applied his theory to the wrong species. Ant societies, of course, are very different from the biped variety, but one common feature is the inherited nature of social behavior. Like any other feature of an organism, behavior can be shaped by evolution. Ants have evolved quite elaborate behaviors, but most are rigidly determined. Human behaviors have also evolved for the purpose of survival. But unlike the programmed instincts of ants, these behaviors, Dr. Wilson believes, are governed by what he calls epigenetic rules, genetically based neural wiring that merely predisposes the brain to favor certain types of action.

In the case of the universal taboo against incest, for instance, Dr. Wilson suggests there is an epigenetic rule that makes people instinctively averse to marrying those whom they knew intimately in childhood.

There are doubtless many other "instinctual algorithms" that guide humans, like other animals, through their life cycle, he says. Altruism, patriotism, status-seeking, territorial expansion and contract formation are among the many behaviors that Dr. Wilson believes are guided by epigenetic rules. "The search for human nature," he writes, "can be viewed as the archeology of the epigenetic rules."

If the essence of human nature is sketched out by the genes, it follows that the major branches of knowledge—sociology, economics, ethics and theology—are shaped by the mind's genetic framework and rest on a foundation of epigenetic rules. The unified tree of knowledge, as envisaged in *Consilience,* has physics at its root, leading to a trunk of chemistry, molecular biology and genetics, and everything else as its branches.

Dr. Wilson criticizes anthropologists and social scientists as having boxed themselves into ideological positions that deny any role for biologically based human nature. He faults economists as "closing off their theory from serious biology and psychology." Hence economics is "still mostly irrelevant" and the esteem that economists enjoy "arises not so much from their record of success as from the fact that business and government have nowhere else to turn."

Ethicists, too, he says, have slipped into the fallacy of ignoring biological origins, in their case the origins of the moral instincts that govern human behavior. Religion, in Dr. Wilson's view, also has epigenetic roots; it evolved from the human equivalent of animals' submissive behavior, and because belief in the supernatural conferred a survival value. Dr. Wilson realizes that he would not now win many votes for this proposition. "The human mind evolved to believe in the gods," he writes. "It did not evolve to believe in biology."

Darwinism is still an affront to many established ways of thought, and *Consilience,* published by Knopf, is not a conciliatory book. It contains enough to disconcert almost everyone. The vigor and aggressiveness of its judgments contrast with the elegance of the writing and the mild and courteous nature of its author. Edward O. Wilson, who is 68, grew up imbued with a respect for social order and civility, but in the course of overcoming personal handicaps and a difficult childhood he acquired an independent mind. "Ed, don't stay on trails when you collect insects," an adviser told him at the outset of his career. "You should walk in a straight line through the forest. Try to go over any barrier you meet. It's hard, but that's the best way to collect."

A straight line to faraway peaks has been Dr. Wilson's path ever since. "I set goals which are usually rather distant and hard to get to, like pointing toward higher mountains," he says. "And I organize my life around that goal."

Much of his life has been devoted to self-imposed endurance tests. At the private school he attended, the Gulf Coast Military Academy, he was drilled in the virtues of hard work and punishingly high standards. In street fights as a boy, he would never quit, obedient to the Academy's ethos, and still bears the scars of his opponents' fists. He learned to turn handicaps to advantage. In his autobiography, *Naturalist,* he describes how the loss of vision in his right eye, stuck by a fish spine, and the loss of hearing for high notes made it hard for him to spot birds. He took up study of insects, which he could still see clearly.

The loss of sight also caused him to be turned down by the Army, a bitter rejection given the respect for military institutions in Southern life. As a boy he was highly religious, reading the Bible twice from cover to cover and being born again as a Southern Baptist. "At my core I am a social conservative, a loyalist," he writes. "I cherish traditional institutions, the more venerable and ritual-laden the better."

A workaholic nature gave Dr. Wilson the impetus to push his quest for scientific truth from one frontier to another. And from some source, perhaps his strong roots in the South and in the cadences of the sermon or the Scripture, he acquired a talent that set him apart from most other scientists—an outstanding gift for language. Two of his books, *On Human Nature* and *The Ants,* a specialist monograph written with Mr. Holldobler, have won Pulitzer Prizes.

Drawn to Harvard by its magnificent ant collection, Dr. Wilson found his outlook and his respect for civility at odds with the "often hard-edged, badly socialized scientists with whom I associate." The Harvard biology department was a battlefield during the 1960's and 1970's between the new breed of molecular biologists and traditionalists who studied whole animals. The molecular biologists, in Dr. Wilson's view, included some of the least socialized of the breed.

Of course it is the molecular biologists' annexationist credo, that nothing in biology makes sense until explained in terms of molecular biology, that Dr. Wilson is seeking to apply to the rest of knowledge. Social scientists and anthropologists are likely to resist the takeover proposed in *Consilience* as strongly as Dr. Wilson and his colleagues asserted their own discipline's integrity against the molecular biologists.

Dr. Wilson has few expectations of instant victory. Noting that defenders of an orthodoxy rarely accept a new idea on its merits, he quotes a remark by the economist Paul Samuelson that knowledge advances "funeral by funeral."

"I don't want to sound Marxist and say it is inevitable," Dr. Wilson says, but he believes that the unification of knowledge will be achieved by a younger generation of scientists who see the opportunities. "By assistant professors, I'm not talking about full professors at Harvard," he says, laughing at the thought of any ray of enlightenment from such a source.

If human nature is shaped by genetic history, then the same techniques now being developed to cure genetic diseases will later be used to change the genes that legislate the contours of the human mind. That capability will raise the question of human purpose in a practical sense: do we wish to change it or leave it as shaped by evolution? Dr. Wilson terms this impending quandary the phase of "volitional evolution."

The prospect, he writes in *Consilience*, will present "the most profound intellectual and ethical choices humanity has ever faced . . . Our childhood having ended, we will hear the true voice of Mephistopheles." His answer is that human nature should be left as it is: "Neutralize the elements of human nature in favor of pure rationality, and the result would be badly constructed, protein-based computers."

The apparent imperfections in human nature, like adolescent violence, may spring from the same epigenetic rule that guides explorers and mountain-climbers, he suggests. Any attempt to change the apparent imperfections in human nature "would lead to the domestication of the human species—we would turn ourselves into lapdogs," he says.

Does he have a new work of synthesis in mind? Dr. Wilson parries the question. He is working full time on his knitting, the monograph about the *Pheidole* genus of ants. "There is a very special pleasure in looking in a microscope and saying I am the first person to see a species that may be millions of years old," he says. "And there you see, somehow I am going to get out of this the answer to the question of why are some species so inordinately successful." Another synthesis seems to be brewing already.

—Nicholas Wade
April 1998

SPEAKING FOR GENES, COME WHAT MAY

Come back with me now to October 1935, to Pensacola, for a walk up Palafox Street. Let's start by peeking over the sea-wall that closes the south end of the street. Down there on the rocks, kept wet and alga-covered by the softly lapping water of the bay, sits a congregation of grapsid crabs.

That's the beguiling voice of Edward O. Wilson, getting ready to tell readers of his autobiography, *Naturalist,* not about grapsid crabs but about the nearly crushing confluence of forces that turned the shy child of divorced parents into a student of nature. Moved to a new school almost every year and seeing his parents only fitfully, Wilson as a boy clung to the only constant feature of the changing scene, the plants and animals of the southern towns where he was raised.

He became an expert on ants and, through an interest in their social systems, wrote the book that first thrust him into public view when it was furiously assailed by a group of left-wing colleagues. *Sociobiology,* published in 1975, looked at the four pinnacles of sociality in evolution, including that of people. Dr. Wilson speculated that, as with other species, there was a genetic basis to many aspects of human behavior, including territoriality, the propensity to live in small groups, male dominance and the ease with which individuals can be indoctrinated into hierarchical systems.

Dr. Wilson's critics, who included several of his Harvard colleagues, said that even if such genes existed—which they doubted—Dr. Wilson's thesis could be used to justify the status quo, the suppression of women and other retrograde politics. They accused Dr. Wilson of espousing "biological determinism," the belief that genes determined a person's fate.

Dr. Wilson responded that his critics had distorted his views and were fallaciously arguing that what is natural is right. In later writings he has made clear that he envisages the genes involved in human behavior as operating in a loose, not dictatorial, style.

Though he stoutly defended his book, he was personally somewhat shaken by the attack, especially as so few of his colleagues came publicly to his defense. Because of the intensity of the attack on sociobiology, other academics have pursued Dr. Wilson's ideas under different labels, such as evolutionary psychology.

In his most recent book, *Consilience,* Dr. Wilson has returned to the idea that human behavior, like that of all other animals, is adaptive—the biologist's word for features that are shaped by the genes. The shaping is so pervasive, in his view, that a wide swath of human institutions, including religion and economics, are genetically based.

Dr. Wilson is perhaps the best-known biologist of his generation. His books are widely read. Although he rarely seeks publicity, he has done so on behalf of the imperiled ecosystems in the world's shrinking tropical forests.

Yes, follow him for a stroll up Palafox street and a peek over that sea wall. He won't throw you to the grapsid crabs. But no one has tried harder to tie you and the crabs and the algae together in a seamless whole, the unity of life.

—N.W.

Andrew Wiles

Quiet Conqueror of a 350-Year-Old Enigma

Norman Y. Lono for The New York Times

Dr. Andrew Wiles is a quiet, diffident man with a shy smile. Yet he summoned the courage to invest seven years working in secret on the world's most famous math problem, a formidable challenge that had baffled the best mathematicians for more than 300 years.

On June 23, 1993, despite all odds, he was able to announce that he had solved it. It was a feat that, if sustained, will catapult the 40-year-old mathematics professor at Princeton University into the ranks of his subject's giants, assuring that the pinnacle of his career has been reached and making him, instantaneously, a sort of icon.

The problem, known as Fermat's last theorem—and familiarly as FLT to its adherents—dates to 1637. The illustrious French mathematician Pierre de Fermat scribbled it in the margin of a book, adding that he had a marvelous proof that, unfortunately, would not fit in the margin. Great and not-so-great mathematicians have been trying to prove it ever since.

Dr. Wiles's proposed proof could still turn out to have a subtle flaw, some mathematicians warn. His 200-page manuscript has not yet been published. And it will take time for experts to check it fully. But unlike the many previous claims, all of which turned out to be flawed, Dr. Wiles's work has been hailed by the leaders of the field.

Already, in Princeton, a town where people routinely drive visitors past the house where Albert Einstein lived, Dr. Wiles's celebrity has begun. Dr. Simon Kochen, chairman of Princeton's mathematics department, said he was receiving congratulations even from strangers.

Dr. Wiles was all but impossible to find in the days after he announced his proof at Cambridge in England, so the thousands of telephone calls, computer messages and faxes from around the world have rained in on others, like Dr. Kochen and other members of Princeton's mathematics department, instead.

"I don't know what it is," Dr. Kochen said. After all, he said, mathematics is practiced by a few thousand professionals and it often seems to be of little interest to the outside world. But with Dr. Wiles's proof, he added: "It's like the human spirit still lives. It's like listening to one of the late Beethoven quartets. It's like we live now and this happened during our lifetimes."

So why Dr. Wiles? And why now? Dr. Wiles, in an interview at his office and at his house a few blocks away, explained that finding the proof was a matter of being in the right place at the right time. And it was a matter of being obsessed with the problem.

Fermat's last theorem is deceptively simple to state. It says that there are no positive whole numbers that solve the equation $x^n + y^n = z^n$ when n is greater than 2. When n is 2, solutions are easy to find. For example, $3^2 + 4^2 = 5^2$. But that, by Fermat's last theorem, is the end of the line. There are no solutions for when n is 3 or any greater number.

Dr. Wiles says he has been fascinated by Fermat's last theorem since childhood. It was the theorem that drew him into mathematics.

The son of a theologian at Oxford University in England, Andrew Wiles first came across Fermat's last theorem when he was 10 years old and saw it in a book from his town's public library. He has forgotten the book's title and author, but he vividly remembers its effect. It made him want to be a mathematician and it made him want to solve the problem.

"I spent much of my teen-age years trying to prove it," Dr. Wiles recalled. "It was always in the back of my mind." But once he became a professional mathematician, he said, he realized that there was a lot more to it than simply working on the problem with a teenager's enthusiasm.

The problem had foiled mathematicians from Fermat's day onward. Fermat himself proved that it was true when the *n*'s, called the exponents in the equations, were 4. A hundred years later, in 1780, the mathematical genius Leonhard Euler proved the theorem was true for the exponent 3, and over the next 50 years others proved it true for the cases of 5, 7 and 13. But Fermat's majestic theorem refers to an infinity of numbers. The one-by-one approach would never prove it.

The turning point came about a decade ago when Dr. Gerhard Frey of the University of the Saarland in Germany suggested that Fermat's last theorem would follow automatically if another, seemingly unrelated, proposition were true. That proposition, known as the Taniyama conjecture, dealt with mathematical objects known as elliptical curves, a class of cubic equations that had been intensively studied by Fermat himself, although not in connection with his theorem.

Then, seven years ago, Dr. Kenneth Ribet, a mathematician at the University of California at Berkeley, formalized Dr. Frey's suggestion, showing that if a crucial part of the Taniyama conjecture could be proved, Fermat's last theorem would be true.

Dr. Ribet said that the Taniyama conjecture was part of a sort of grand unified theory of mathematics. "The Taniyama conjecture is part of the vast Langlands philosophy," Dr. Ribet said. He explained that Dr. Robert Langlands, a professor at the Institute for Advanced Study in Princeton, suggested years ago that two apparently disparate fields of mathematics were actually one and the same.

The fields are algebra, which deals with equations, and analysis, which, like calculus, deals with how mathematical structures vary around the neighborhood of a point on a surface. The Langlands philosophy, Dr. Ribet said, "is a deep, far-reaching vision in mathematics." The Taniyama conjecture, he added, "is a special case of what Langlands suspected."

Dr. Ribet's proof linking the conjecture to Fermat's last theorem fired Dr. Wiles's imagination. It said, essentially, that you could associate an elliptic curve with a solution to the Fermat equations. If you could just show that the elliptic curve could not exist, then you could show that neither do solutions to the Fermat equations.

"I knew when I heard about Frey and Ribet's result that the landscape had changed," Dr. Wiles said. "There is no question that what they did completely changed the problem for me psychologically. Hitherto, Fermat's last theorem was like other problems in number theory. It would go on for thousands of years and never be solved and mathematics would hardly notice."

But, Dr. Wiles added: "What Frey and Ribet did was make Fermat's last theorem a consequence of a problem that mathematics could not ignore. Too much depended on it. A whole architecture of future problems depended on it. For me, that meant that the problem was going to be solved. And once I had that confidence, I couldn't let go."

The very day that Dr. Wiles heard about Dr. Ribet's result, he dedicated his life to using it to prove Fermat's last theorem.

He worked at home on his secret project, sitting in a barren attic office on the third floor of his Tudor-style house. No computer was necessary, and no telephone was present to intrude on the absolute silence.

"If he said he was working on Fermat's last theorem, people would look askance," Dr. Kochen said. "And if you start telling people who are experts, you end up collaborating with them. He wanted to do it on his own." Only one person in the Princeton mathematics department knew what was going on: Dr. Nicholas Katz, who agreed to serve as sort of a sounding board for Dr. Wiles. But Dr. Katz was sworn to secrecy.

"I made progress in the first few years," Dr. Wiles said. "I developed a coherent strategy." Then, he said, two years ago, he had a pivotal idea that reduced the problem to a calculation. The calculation was "one that was similar to one others had tried and hadn't achieved, but it was a problem that was doable," he said.

Dr. Wiles worked feverishly on it. "Basically, I restricted myself to my work and my family," he said. "I don't think I ever stopped working on it. It was on my mind all the time. Once you're really desperate to find the answer to something, you can't let go."

Eventually, Dr. Wiles said, he had solved all but one critical case for his proof. The final barrier fell not long ago when he was reading a paper by Dr. Barry Mazur, a mathematician at Harvard University. "I can't even remember why I was looking at it," Dr. Wiles said. But he took notice of the mention of a certain construction. "It was immediately clear to me that

I should use this construction and that I would solve this last problem I had."

The construction, as it happened, was not new. It dates to the 19th century. But Dr. Wiles had not heard of it before. As soon as he saw it, Dr. Wiles said, "I knew I had it." He could complete his seven-year quest and the proof of Fermat's last theorem.

Dr. Wiles presented his work in three lectures at a small mathematics meeting at Cambridge University in England.

Unlike many scientists who trumpet every tiny advance in papers and press release, Dr. Wiles gave no hint of the bombshell he was about to detonate over the world of mathematics. The audience was not even certain he would mention Fermat, though some had their suspicions from the drift of the argument. Between lectures, e-mail messages started flashing between England and the United States as mathematicians speculated on his target.

One reported: "Andrew gave his first talk today. He did not announce a proof of Taniyama-Weil but he is moving in that direction and he still has two more lectures. He is still being very secretive about the final result."

Dr. Kochen, waiting back in Princeton, was tense with anticipation. He and another Princeton mathematician, Dr. John Conway, could talk of nothing but whether the proof would be accepted by the handful of mathematicians who understood the field.

"Tuesday night, neither of us could sleep," Dr. Kochen said. Around 5:30 Wednesday morning, Dr. Conway unlocked the door to Fine Hall, the brown tower that houses the mathematics department, went into his office, and turned on his computer. The first message arrived at 5:53 A.M. from Dr. John McKay of Concordia University in Montreal, who was attending the meeting. "F.L.T. proved by Wiles," it said.

Not until the end of the third and last lecture had Dr. Wiles cited the proof in which his argument culminated.

Dr. Wiles returned home, elated but tired, and spent time with his wife and young daughters. "I've been to the swings," he said, and to a party given by friends. He has tried to absorb the consequences of his achievement.

It is also a release to have completed the proof. "I haven't let go of this problem for nearly seven years. It's gone on on a day-to-day basis. I've almost forgotten the experience of getting up and thinking about something else," he said.

Yet the release is accompanied by a bittersweet feeling at seeing this most famous of all mathematics problems vanish.

There is a certain sadness in solving the last theorem, Dr. Wiles said. "All number theorists, deep down, feel that," he said. "For many of us, this problem drew us in and we always considered it something you dream about but never actually do." Now, he said, "There is a sense of loss, actually."

—GINA KOLATA
June 1993

WHAT'S HIGHER THAN EVEREST?

What do you do for an encore after you stun the world by solving one of history's most celebrated math problems? If you're Dr. Andrew Wiles of Princeton University, you go back into seclusion and you don't tell anyone what you are doing.

Of course, Dr. Wiles does not really have to achieve anything else for the rest of his life. After being forced to concede that there was a gap in the proof he had first announced, Dr. Wiles in 1994 was able to offer a complete solution for Fermat's last theorem, reached with the help of a former student, Dr. Richard Taylor of Cambridge University. The story of Fermat's last theorem and its proof became the subject of a popular book, a public television show and numerous lectures at mathematics meetings around the world. The feat, which earned Dr. Wiles numerous awards, is clearly one for the history books, more than enough for a career.

Dr. Wiles could become like Albert Einstein. "Here was a man who had perhaps the greatest intellect that God ever created, and, in the last years of his life, nothing much came of it," was the assessment of Dr. Herbert Robbins, a statistician who knew Einstein when he was at the Institute for Advanced Study at Princeton.

In an interview published in 1985, Dr. Robbins said Einstein "never complained about it, and no one mentioned it, but everyone knew he was essentially finished as a scientist." The same thing happened with Isaac Newton. "From about age 30 on, he did absolutely nothing in science," Dr. Robbins said.

But few think Dr. Wiles, who turns 47 in 2000, is finished with mathematics. From remarks he has made and hints he has dropped, many think he is working on a broader problem that his proof of Fermat's theorem opened up. That problem, sometimes known as the Langlands conjecture, could unify disparate areas of mathematics. It is sort of the mathematicians' equivalent of the physicists' grand unified field theory.

But whatever Dr. Wiles is doing, others in his field say he has been aloof. For example, Dr. Kenneth Ribet, a mathematician at the University of California at Berkeley, did much of the pathbreaking research that convinced Dr. Wiles in 1986 that he should attempt his proof of Fermat's last theorem. It was Dr. Ribet's results that inspired Dr. Wiles to seclude himself in his attic in his Tudor home near the leafy Princeton campus and work, alone, for seven years on his proof.

Dr. Ribet said he invited Dr. Wiles to speak at a major mathematics conference in December of 1999 on the extensions of his work on Fermat's last theorem. But Dr. Wiles did not come.

"He declined, saying he was too busy," Dr. Ribet said. "He is an extremely secretive individual. I have had no idle communications with him since 1993."

Even fellow mathematicians at Princeton say they are in the dark about what Dr. Wiles is up to.

But then again, said one colleague, Dr. John Conway, mathematicians often hesitate to burst in on their peers, like ancient mariners out of Coleridge's poem, and regale them with developing research.

"It's always an imposition when you do that," Dr. Conway said. "The listener has to pretend to be interested." The reason, he explained, is that these complex arguments can be excruciatingly difficult to follow, requiring that the listener think hard and spend a lot of time trying to understand the truth of the logical arguments. Mathematicians often fear—and rightly so, Dr. Conway says—that their colleagues will listen with an air more of forbearance than enthusiasm.

"A tells B something. B doesn't really understand it but sort of suspends disbelief for the sake of the friendship," Dr. Conway said.

So some, at least, are just as happy when Dr. Wiles keeps his ongoing research to himself.

—G.K.

Temple Grandin

Empathy for Animals, Through the Lens of Autism

Brian Gadberry for The New York Times

"Cattle are a prey species, so they're very vigilant," said Dr. Temple Grandin, hunching over to get a cow's-eye view of the curving chute of a feedlot she designed for 30,000 cattle here in Brush, Colorado. "If something in their environment looks like it shouldn't be there"—she picked up a Styrofoam cup tossed to the ground—"they won't go in."

"When a plant is working correctly, it's free of distractions, like chains jingling or air blowing in their faces," said Dr. Grandin, who has a doctorate in animal science from the University of Illinois. She also has autism, which has given her a window into the animal world that most people can

Design for Slaughter

By looking at things from a cow's perspective, the animal scientist Temple Grandin found simple ways to make slaughter less traumatic for animals and safer for human handlers. She says her autism enables her to visualize the way her machines will work before they are built. Today most cattle in the United States and Canada are handled in facilities she designed.

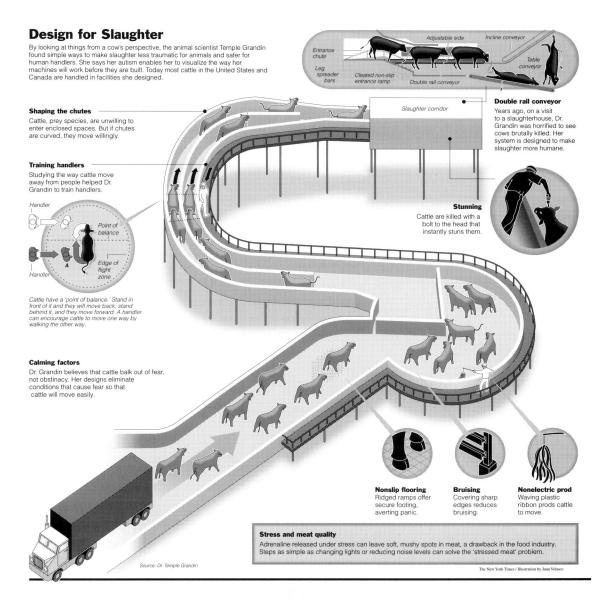

Shaping the chutes
Cattle, prey species, are unwilling to enter enclosed spaces. But if chutes are curved, they move willingly.

Training handlers
Studying the way cattle move away from people helped Dr. Grandin to train handlers.

Handler
Point of balance
Handler
A
Edge of flight zone

Cattle have a 'point of balance.' Stand in front of it and they will move back; stand behind it, and they move forward. A handler can encourage cattle to move one way by walking the other way.

Calming factors
Dr. Grandin believes that cattle balk out of fear, not obstinacy. Her designs eliminate conditions that cause fear so that cattle will move easily.

Entrance chute *Adjustable side* *Incline conveyor* *Table conveyor* *Leg spreader bars* *Cleated non-slip entrance ramp* *Double rail conveyor*

Slaughter corridor

Double rail conveyor
Years ago, on a visit to a slaughterhouse, Dr. Grandin was horrified to see cows brutally killed. Her system is designed to make slaughter more humane.

Stunning
Cattle are killed with a bolt to the head that instantly stuns them.

Nonslip flooring
Ridged ramps offer secure footing, averting panic.

Bruising
Covering sharp edges reduces bruising.

Nonelectric prod
Waving plastic ribbon prods cattle to move.

Stress and meat quality
Adrenaline released under stress can leave soft, mushy spots in meat, a drawback in the food industry. Steps as simple as changing lights or reducing noise levels can solve the 'stressed meat' problem.

Source: Dr. Temple Grandin The New York Times / Illustration by Juan Velasco

hardly imagine. "If a little chain is hanging at the entrance of the chute, that leader animal will just stop and follow the movement of the chain."

Dr. Grandin, tall and lanky in her jeans and cowboy boots, stopped in her tracks, moving her head back and forth the way a cow does, watching a moving chain. Then she stepped ahead, carefully, one foot at a time. "It could be as simple as a shadow across the entrance," she said, "or a coat on a fence, or seeing a person through the slats."

Two students who are working with Dr. Grandin, who teaches part time at Colorado State University, followed closely behind, pointing out cow-like perceptions to a stranger. "If this were a straight shot to the truck, you wouldn't go, it's too scary," said Mark Deesing, a horse trainer with ideas of his own about animal behavior. "Circles are good because there's nothing they can see, but there's still a place to go. It's important it doesn't look like a dead end."

The sorting pens are set on a diagonal, to follow the flow of the herd, and have no right angles to bruise the animals. "The sides of the chute are solid, so you can't see anyone on the other side," said Jennifer Lanier, a former zookeeper who now works with bison as a doctoral student at Colorado State.

At 50, Dr. Grandin has designed livestock facilities for nearly half the cattle in the United States and Canada, from the feedyard, where animals spend four months to a year beefing up, to the slaughterhouse, where they move along what Dr. Grandin calls the "Stairway to Heaven," to be killed instantly by a metal bolt through the head.

Dr. Grandin has devoted her life to the humane treatment of animals, and her affinity for them is more visceral than intellectual.

"I think in pictures, like an animal," she had said at breakfast that morning. "My nervous system is more like an animal's. The sounds that bother me are the same sounds that bother an animal. My emotions are simple—and the main one is fear."

Autism, which affects 400,000 Americans, is a neurological disorder that leaves some people unable to speak or function and others with savant-like talents for things like drawing, music or mathematics. It may be genetically based.

Like many autistics, Dr. Grandin says she is oblivious to the little nuances that usually flow between people. "I didn't even know about eye signals, until I read about it two years ago," she said, and she is completely baffled by romantic love.

She speaks fast, in a voice that is a bit loud (she has worked for years to tone it down) and she often repeats her thoughts, word for word (that, too, was worse when she was an adolescent, and her classmates called her "tape recorder"). Her pale blue eyes do not hold a gaze.

"It's too distracting to look at a face," she said. "Any subtle cue I do get is from tone of voice."

"I troll through the Web page of my mind and I have no emotions," she said. "I can hold them there, like when my aunt died, but I'm just surfing the Web of my mind. My emotions are not seamlessly lined up in my database. I have to work on pictures in my imagination for about five minutes and then I'll get worked up about it."

Like the prey animals she works with, Dr. Grandin is hypersensitive to sound and touch and has a kind of fearful readiness to flee. One night in a hotel bed, for example, she woke, heart pounding, at the sound of a truck with a backup alarm 23 floors down.

Antidepressants have muffled this "massive fear response," she said. "But the orienting response is still there, like a deer turning its head if it hears a funny noise. Is it going to keep grazing, or run away?"

Scientists attribute that fear response to abnormalities in the brain, including the amygdala, which registers fear, and the fight-or-flight response.

"The cerebellum acts as the modulator for the senses," said Dr. Grandin, citing a 1994 study by a Boston pediatrician, Dr. Margaret Bauman, whose brain autopsies on autistics indicated immature development of the cerebellum and limbic system.

As an infant, Temple was so sensitive to touch that she would scream and stiffen when her mother tried to hold her. She could understand speech, but could not form words. And like many autistic children, she loved to rock—to block out the noises assaulting her hypersensitive system.

At 2½, she was labeled brain-damaged and doctors recommended institutionalization. But Temple's mother enrolled her in a nursery school that specialized in speech therapy, and hired a nanny who kept her occupied with creative games. And the whole family sat down together for meals.

"I couldn't sit and rock if I had to pass the meat and potatoes," said Dr. Grandin. "It was 40 hours a week of being tuned in."

When Temple was in second grade, she started dreaming of a "hug" machine that would exert steady pressure on her body without overwhelming her hypersensitive nervous system. Later, at an aunt's ranch, she saw how the cattle reacted to the squeeze chute, a metal device that holds animals in place for vaccinations and other procedures, and recognized it as a crude version of her dream of a hug machine. She talked her aunt into letting her use it, and felt her anxiety lessen as the metal sides of the chute pressed against her body.

"At 18, I built my first device," she said, encouraged by a high school science teacher, who showed her how to use medical abstracts to research why the machine calmed her.

Since then, she has improved the squeeze machine, with inflatable, cushioned sides, a headrest and hydraulic controls, and she keeps one by her bed at home, to reduce anxiety and to relax. The patented machine is widely used by autistic adults and children, especially in schools.

"We have youngsters who will ask for it," said Lorna Jean King, an occupational therapist and director of the Center for Neurodevelopment, in Phoenix, Arizona. "When they get nervous, they'll say, 'I need the hug machine.' Autistic kids are almost always hyper-responsive and don't want to be touched. But sustained pressure, over time, damps that down."

In 1986, Dr. Grandin opened a window on the poorly understood world of autism with her first book, *Emergence: Labeled Autistic,* and followed it about a decade later with *Thinking in Pictures.* Both books illuminated not only the problems of autism, but its gifts—in her case, heightened powers of concentration and almost uncanny visual abilities.

"I can test-run equipment in my head," she said. "I rotate an object in space. I walk around it. I can take an aerial view. I fly over a design in my mind. I walk through it. Virtual reality doesn't get me very excited. I can walk through a set of plans in my mind."

Early in her career, Dr. Grandin was fired from a job when she argued with engineers over a design that moved beef carcasses off a conveyor, via a chain hooked to an overhead track. She could see from the drawings that it would not work, but no one listened. On the first run-through, the track was pulled out of the ceiling. "They couldn't visualize," she said of her colleagues.

About 25 years ago, when Dr. Grandin was working on her master's thesis at Arizona State University, she went through the chutes and the corrals, at cow height, snapping pictures of everything a cow might see, trying to figure out why they balked at one chute and walked easily into another. She put herself in the cow's place, seeing shadows falling through a slatted fence, jumping at a hissing air valve, imagining the panic of hooves slipping on a metal ramp.

When she was asked to design a dip vat at what was then John Wayne's Red River feed yard in Arizona, she walked through the feedlot. A dip vat is a long, narrow, deep pool, which is filled with a pesticide that kills parasites on animals. The old design forced cattle to slide into the vat down a steep, slick concrete slope. "Imagine a whole bunch of people piling up behind you in front of an airplane slide into the ocean," said Dr. Grandin. "You'd panic."

So she replaced it with a concrete ramp with deep grooves, which appeared to enter the water gradually, but in fact, dropped off abruptly into deep water. She tested it in her mind and saw the cows stepping out over the water and falling in.

"But the cowboys didn't think it would work," said Dr. Grandin. "That's the way they think, that you have to force cattle to do everything."

When she was not around, workers put a metal sheet over the ridged ramp—and two animals drowned because they panicked on the slide and flipped over on their backs. When Dr. Grandin insisted the sheet be removed, the slide worked perfectly. She calls the design "cattle walking on water."

She has no formal training in engineering or drafting. She absorbed it, almost by osmosis, by spending a lot of time watching an engineer draft designs at one of her first jobs, with a feedlot construction company.

Her ability to think in pictures, coupled with her empathy for animals and a near superhuman energy, has made Dr. Grandin a unique resource for not only the meat industry, but horse trainers, zookeepers and bison ranchers. She has trained antelopes at the Denver Zoo to stand calmly in a box she designed, while blood samples are taken. And she is working with her two students on radical new ways to train and care for wild or highly excitable animals without using force.

"We've got to get rid of the cowboy rodeo stuff, the yelling and scream-ing at them," said Dr. Grandin, who is convinced that if animals are unco-operative, it is out of fear, not obstinacy.

Dr. Grandin has a combination of knowledge, no-nonsense objectivity and irrepressible individuality (like wearing jeans and cowboy boots to black-tie dinners) that sits well with the cattle producers.

"She gets in there and cusses with the best of us," said Dr. Gary Cow-man, an executive at the National Beef Cattlemen's Association in Engle-wood, Colo. "Sometimes somebody won't agree with her, but she's highly respected as the authority in animal care and behavior in the world."

She has brought "a very sophisticated science element to animal agri-culture," he said. "She's done in-depth evaluations of animal behavior. She's worked on beef cattle handling facilities, running them through the chute for their vaccinations. She's studied their sight, and how they move. And then she has the ability to develop facilities that minimize their fear and stress. Without a doubt, over the past 20 years, her influence has been out there."

Meatpackers call on her, too. "When you say humane handling in packing plants, you say Temple Grandin," said Janet Riley, who heads the animal welfare committee for the American Meat Institute, a trade association in Arlington, Virginia. Many of the group's 400 members have adopted Dr. Grandin's guidelines for running a humane operation, she said, and because the animals are calmer, the plant runs more efficiently.

"And the product is better," said Ms. Riley. "If an animal is stressed, adrenaline will be released, and you get some soft, mushy spots in the meat."

Dr. Grandin is often called in to troubleshoot for a plant with "stressed meat"—like the one that was cleaning its pens at night with a front-end loader with a beeping backup alarm. "I spent a few minutes on the night shift and said, 'Get that thing out of there,' " Dr. Grandin said.

But the changes she is most proud of are the radical redesigns of systems that can make death painless for the animals.

In her book, she describes the first shackling and hoisting system she saw, 17 years ago, in a now-defunct kosher meat plant in Spencer, Iowa: "Employees wearing football helmets attached a nose tong to the nose of a writhing beast suspended by a chain wrapped around one back leg. As I watched this nightmare, I thought, 'This should not be happening in a civilized society.' In my diary I wrote, 'If hell exists, I am in it.' I vowed that I would replace the plant from hell with a kinder and gentler system."

Kosher slaughterhouses are exempt from the Humane Slaughter Act, which outlaws shackling and hoisting systems, but Dr. Grandin has ripped out half a dozen of them for plants that have changed voluntarily.

Years ago, Dr. Grandin invented a system for slaughterhouses in which each animal is guided onto a double-rail conveyor, which it straddles, supported under the belly and chest, much as a person rides a horse. Solid walls on either side of the conveyor keep the animal stable, and in close connection with its fellows, comfortingly nose to rump, much the way animals move single file in a pasture. For kosher slaughter, each animal enters a restraining box, with a head-holding device that allows the rabbi to perform the ritual cut, without pain or fear to the animal.

When a kosher system was installed in Alabama a few years ago, Dr. Grandin operated the hydraulic gears of the restraining box herself, concentrating on easing the animal into the box as gently as possible.

"When I held his head in the yoke, I imagined placing my hands on his forehead and under his chin and gently easing him into position," she wrote in her book. "Body boundaries seemed to disappear, and I had no awareness of pushing the levers."

She compared it to a state of Zen meditation: "The more gently I was able to hold the animal with the apparatus, the more peaceful I felt. As the life force left the animal, I had deep religious feelings. For the first time in my life logic had been completely overwhelmed by feelings I did not know I had."

A slaughterhouse certainly "makes you look at your own mortality," said Dr. Grandin. "Those animals just walk into the chute and it's all over. If you see them cut up, they're so fragile inside. People are made of the same stuff."

She designed her first Stairway to Heaven at the former Swift plant in Tolleson, Arizona, which is also where she killed her first animal. "When I got home, I couldn't believe I had done it," she said. "It was very exciting. I was scared that I'd miss, because it does take some skill, and I knew that I could hurt the animal. I couldn't bring myself to use the word, 'kill.' "

Early this spring, Dr. Grandin completed a survey of 24 slaughterhouses across the country for the United States Department of Agriculture, rating each one for humane treatment and efficiency. The solutions she suggested were often as simple as training employees how to herd the animals without excessive electrical prodding, installing nonslip grates on a slippery floor, rotating jobs, and the proper maintenance of stunning guns and electrical stunners so that animals suffer no pain.

Of the 24 plants, "five were excellent and five were terrible," she said. "The single most important thing is the attitude of the management. If you have management that cares, the animals will be handled nicely. With the bad ones, management thinks poking cattle with electric prods is just normal. The good news is, everything that was wrong in that survey is easy to fix."

Dr. Grandin is not apologetic about savoring her steak at the Lone Star Restaurant in Fort Collins.

"I believe that we can use animals ethically for food, but we've got to treat them right," she said. "None of these cattle would have existed, if we hadn't bred them. We owe them a decent life—and a painless death. They're living, feeling things. They're not posts, or machines."

And anyone who drinks milk or eats cheese should know that "a cow has to have a calf every year to keep producing milk," she said, "and those calves are raised for beef."

—ANNE RAVER
August 1997

THE FEAR PRINCIPLE

Dr. Temple Grandin continues to traverse the world at an almost inhuman pace, inspecting cattle feedlots and slaughterhouses, studying other animals like antelope and giving lectures on her research and what it means to be autistic.

Dr. Grandin and two associates, Mark Deesing, a horse trainer who now designs cattle facilities, and Jennifer Lanier, one of her doctoral students at Colorado State University, are working together to revolutionize the handling of wild animals and high-strung horses. Many of the principles Dr. Grandin applies to cattle relate to the handling of wild animals. Simply paying attention to the animal's flight zone allows zoo workers to handle antelopes or ranchers to herd bison into an auction lot without whips or electric prods.

It all goes back to understanding fear, said Dr. Grandin, be it an autistic child flinching at a loud noise, or a bison going berserk at the Denver stock show when a neck restraint gets shoved in its face.

"You can't do that with wild animals," Dr. Grandin said. "You have to get them habituated to the sound of the squeeze chute, or the door being slammed. New experiences are both scary and interesting. If you take a beach towel and throw it in the pasture, every buffalo in the pasture will come and look at it, but if you shove it in his face, he will blow up."

In other research that has great implications for working with animals, Dr. Grandin and her colleagues have linked hair whorls—spirals on the forehead of an animal—to temperament. Animals with high whorls—above the eyes rather than below—are much more likely to be fearful of new situations or rough handling, Dr. Grandin says, and must be trained and handled with gentle patience. Fear also affects the quality and tenderness of meat by raising adrenaline levels and making weight loss and injury more likely.

It was Mr. Deesing who first made the connection of high hair whorls to high-strung horses. He made the rounds of animal behaviorists at the University of Utah, Texas A&M, Cornell and the like, hoping someone would study his hunch, but no academic would listen to a guy who never finished high school.

"Then I met Temple," Mr. Deesing said. "And she said, 'I don't know what the hell you're talking about, but I'm sure you're seeing something.' " Dr. Grandin sent Mr. Deesing off to some of the world's top racetracks and show rings to record the position of the horses' whorls. At Spruce Meadows, a show ring in Alberta, Canada, for example, he found that 22 percent had high whorls, compared to 5 percent of the general population. Eventually, they observed thousands of cattle and bison in stockyards and auction rings.

"It seems that the higher whorls are pushed higher as the brain develops," Dr. Grandin said. "Double whorls are more correlated with race horses. They're more temperamental. Nervousness is one of their traits."

Such knowledge could radically change how horses, especially wild ones, are handled and trained. The Bureau of Land Management recently asked Dr. Grandin to evaluate its wild horse program.

"It can get pretty nasty," said Mr. Deesing, who recently observed the gathering of wild horses in Wyoming. "They get crowded into a little portable round-up pen, where the fighting and biting gets crazy, and then chased up a single-file chute into a truck, where they fight and stampede each other." In one gathering of 75 horses, Mr. Deesing noted that 90 percent had head injuries. He and Dr. Grandin then worked with the same number of wild horses, with almost no injuries. If funded, they plan to evaluate 150 management areas in the country, and to recommend methods for better handling.

Dr. Grandin is also inspecting beef plants for the McDonald's Corporation, and she is heartened by more humane practices. "People are putting away the electric cattle prods," she said.

Dr. Grandin and Mr. Deesing explain much of their recent research on a Web site, www.grandin.com. The Web site contains design drawings and color photographs of animals like sheep calmly circling in the opposite direction of their handlers in a corral, or pigs curiously following the twists and turns of a serpentine chute.

It seems the perfect medium for Dr. Grandin, who compares her thought processes to a computer. It makes the complex principles of animal training as clear as "thinking in pictures," which is how, she says, animals—and people with autism—perceive the world.

—A.R.

Hans A. Bethe

He Lit Nuclear Fire;
Now He Would Douse It

Michael Okoniewski *for* The New York Times

"For the things I do, it's accurate enough," Dr. Hans A. Bethe said as he rummaged through his briefcase and pulled out a slide rule, a relic from the days before computers took over tedious number-crunching for most scientists. Its battered case told of considerable use.

What Dr. Bethe does at the age of 90, and has done for more than seven decades, is ponder such riddles of nature as how stars live and die. It is his passion. Once it won him a Nobel Prize in physics and now it keeps him excited and in his office at Cornell University in Ithaca, New York, where he arrived more than 60 years ago after fleeing Nazi Germany.

A combination lock on a metal cabinet hints at what else he does, his sideline, as he puts it, an avocation of more than a half century that helped change history. The atomic bomb.

Dr. Bethe knows how it lives—having overseen its birth during World War II, having felt its blistering heat across miles of desert sand, having watched its progeny fill superpower arsenals—and now he is working hard to make it die.

He wrote a letter to President Clinton that some advocates of arms control regard as historic. As the most senior of the living scientists who begat the atomic age, Dr. Bethe called on the United States to declare that it would forgo all work to devise new kinds of weapons of mass destruction.

But his dream, it turns out, is much larger than that. In an interview, Dr. Bethe said that a concerted push by the world's nations and peoples might yet cut nuclear arsenals down from their current levels of thousands of arms to perhaps 100 in the East, 100 in the West and a few in between.

"Then," added this survivor of Hitler and Mussolini, his voice gentle but words sharp, "even if statesmen go crazy again, as they used to be, the use of these weapons will not destroy civilization."

Eventually, perhaps late next century, Dr. Bethe said, the right social conditions may finally arise so that the bomb is no more, so that no nation on Earth will want to wield the threat of nuclear annihilation. The nightmare will be over.

He paused. "That is my hope," he said. "My fear is that we stay where we are," with each side keeping thousands of nuclear arms poised to fly at a moment's notice. "And if we stay where we are, then additional countries will get nuclear weapons" and the Earth may yet blaze with thermonuclear fire, the kind that powers stars and destroys most everything in its path.

Hans Albrecht Bethe (pronounced BAY-ta) was born on July 2, 1906, in Strasbourg, Alsace-Lorraine. His father, a physiologist at the university there, was Protestant and his mother Jewish. Hans was their only child.

Displaying an early genius for mathematics, he excelled in school and received a Ph.D. in physics in 1928 at the University of Munich, graduating summa cum laude. He fled Germany after Hitler came to power, going first to England and then to America, arriving at Cornell in 1935.

While helping to found the field of atomic physics, he became fascinated by nature's extremes. In 1938 he penned the equations that explain how the Sun shines and how stars in the prime of life feed their nuclear fires. In 1967 he won a Nobel Prize for the discovery.

From 1943 to 1945 he headed the theoretical division of Los Alamos, the top-secret laboratory in New Mexico where thousands of scientists and technicians, fearful that Hitler might do it first, labored day and night to unlock the atom's power.

Dr. Bethe coaxed some of world's brightest and most idiosyncratic experts to success as they toiled behind rows of barbed wire. Their atomic bomb shook the New Mexican desert on July 16, 1945. The next month the American military dropped similar ones on the Japanese cities of Hiroshima and Nagasaki.

After the war, Dr. Bethe devoted himself not only to nuclear science but to the social dangers posed by that knowledge, in particular to keeping the bomb from ever killing people again.

He advised the Federal Government on matters of weapons and arms limitation, becoming a prime mover behind the first East-West arms accord, the 1963 Limited Test Ban Treaty, which ended nuclear explosions in the atmosphere and permitted them only beneath the Earth.

That stopped the rain of radioactive fallout that had raised the risk of cancer and birth defects among many people. But Dr. Bethe wanted more. He campaigned for a complete cessation to all testing, in opposition to Pentagon planners and politicians intent on redoubling the size of the nation's nuclear arsenal.

The development of new types of nuclear arms requires numerous test firings and, as flaws inevitably come to light, design improvements. The absence of explosive testing sharply increases the odds of failure and virtually rules out the possibility of perfecting new designs.

In the 1980's, Dr. Bethe was on the losing side of the political war over nuclear-arms development as the Reagan Administration pressed ahead with dozens of underground explosions. One series aimed at perfecting a new generation of bombs that fired deadly beams.

In the 1990's, he was on the winning side as President Clinton signed, and the United Nations endorsed, the Comprehensive Test Ban Treaty. Its goal is to halt the development of new weapons of mass destruction by imposing a global ban on nuclear detonations.

A remaining trouble, as Dr. Bethe sees it, is that the United States over the decades has become so good at designing nuclear arms that it still might make progress despite the ban. Indeed, the Clinton Administration recently began a $4-billion-a-year program of bomb maintenance that is endowing the weapons laboratories with all kinds of new tools and test equipment, including a $2.2 billion laser the size of the Rose Bowl that is to ignite tiny thermonuclear explosions.

Critics fear the custodians might get carried away, begetting new designs and perhaps even new classes of nuclear arms.

So it was that Dr. Bethe wrote President Clinton in April, asking for a pledge of no new weapons. "The time has come for our nation to declare that it is not working, in any way, to develop further weapons of mass destruction," he wrote.

The United States "needs no more," Dr. Bethe stressed. "Further, it is our own splendid weapons laboratories that are, by far and without question, the most likely to succeed in such nuclear inventions. Since any new types of weapons would, in time, spread to others and present a threat to us, it is logical for us not to pioneer further in this field."

In the interview, Dr. Bethe waxed philosophic about the odds that his personal appeal might engender new Federal policy. "It's a big step for the President to say so, but it's a small step for me," he mused. "Maybe the laboratories will feel that my letter was useful and maybe they'll even follow my advice. I think that's all one can expect."

The issue is important, he added. If the community of nations comes to view the United States as a nuclear hypocrite, whether correctly or not, that perception could threaten to undermine the new treaty and its ratification around the world. Instead, Dr. Bethe said, the United States must been seen as striving to obey the letter of the law.

Dr. Bethe's face comes alive as the topic turns to his current scientific research: how a single aging star can suddenly explode with the power and brilliance of an entire galaxy of 100 billion stars.

It seems like pure poetry given the light he himself is now shedding in his final years.

"I want to understand just how the mechanism works," Dr. Bethe said, "how you get a shock wave that propels most of the star outward, propels it at very high speed."

Most days, he said, he spends about four hours studying the nature of the exploding stars, which are known as supernovas. Occasionally, he works up to six hours.

Theoretic physics is a quintessential young man's field, where geniuses often peak at the age of 30, like athletes. Very few make significant contributions at 50. But at 90, Dr. Bethe, a living legend among his peers, is still going strong. "Here's my latest paper," he said with a grin, displaying it proudly on his cluttered desk. "It has been accepted by *The Astrophysical Journal*." The main point, he said, "is that it's easy to get the supernova to expel the outside material," eliminating the problems theorists once encountered.

Dr. Bethe is not interrupting his research to write memoirs. Instead, a biographer is at work. "It's much easier to have a biographer," he remarked, "and he writes much better than I do."

The back of his office door, in an easy-to-view position, held a poster of the Matterhorn. For nearly a half century, a small town at the foot of the great Swiss mountain has been a vacation spot for Dr. Bethe and his wife, Rose Ewald, whom he met in Germany and married in 1939 while the two were newcomers to the United States.

"I couldn't live without her," he said.

His hair askew, his eyes agleam, Dr. Bethe looked a bit like an aged wizard on the verge of disappearing in a puff of smoke. He seemed at ease with his many lives over many decades and appeared to have reconciled his early work on the bomb with his current push to eliminate it. For him, doing the right thing in different periods of history seemed to call for different kinds of actions.

"I am a very happy person," he said with a relaxed smile. "I wouldn't want to change what I did during my life."

—WILLIAM J. BROAD
June 1997

DREAM OF A TEST BAN DIMS

A distinguished rebel of the atomic establishment, Dr. Hans Bethe fought hard throughout the 1990's for nuclear disarmament but saw his goal suffer a serious blow when in October 1999 the Senate rejected the nuclear test ban treaty. Despite that, and despite the lengthening shadow of his years, Dr. Bethe continued to bring his calm, reasoned voice to the atomic debate, still hoping to sway opinion and douse the fire he helped ignite.

The Comprehensive Test Ban Treaty, a global accord meant to end the development of new kinds of nuclear arms, was for decades the holy grail of arms controllers. President Dwight D. Eisenhower proposed it in 1958 as a way to tame the world's nuclear fires and slow a tense arms race with the Soviet Union. Just as new jets cannot be made without test flights, new kinds of nuclear arms usually must be detonated repeatedly so that flaws may be spotted and potency ensured.

The treaty goal lay dormant for most of the Cold War, but was resurrected and pushed strongly by President Bill Clinton. In 1995, he called for negotiations for a permanent global ban on all nuclear testing and in 1996 became the first head of state to sign the resulting accord, followed by 151 other nations. Mr. Clinton called the treaty "the longest-sought, hardest-fought prize in the history of arms control."

As the Senate prepared to pass judgment in late 1999, Dr. Bethe joined with 31 other Nobel laureates in physics to urge approval, calling the treaty "central to future efforts to halt the spread of nuclear weapons." It was the first time so many prominent American physicists had shown such public unity.

But after sharp debate, the Republican-led Senate handed the Democratic President a stinging rebuke on October 13, 1999, voting down the accord 51 to 48. "The founding fathers," said Senator Trent Lott, the majority leader and a foe of the agreement, "never envisioned that the Senate would be a rubber stamp for a flawed treaty."

Dr. Bethe was outraged. "The Senate's rebuff," he wrote in the November 18 issue of *The New York Review of Books,* "made me sad, both for my country and the world" because the accord was good in itself and because the ratification failure "will have serious consequences for American foreign policy."

The ban, he said, worked strongly in favor of the United States, since Washington would keep its huge advantage in nuclear arms while making the emergence of new powers "extremely difficult." Most criticisms of the treaty, he added, were fatuous. It could be policed for militarily significant cheating and would cause America's own arsenal no troubles.

"If any component shows signs of deterioration, it is refabricated," Dr. Bethe wrote. "It is well established that, with existing apparatus, the efficacy of the refabricated units can be proved by sophisticated non-nuclear tests. We do not need explosive tests to prove that these weapons work."

The treaty, though damaged by the Senate rejection, is not considered dead. For years, nations have done no known explosive testing of nuclear arms, and treaty proponents vowed to keep the issue alive so the accord in time might get another day in the Senate, and the world another chance to deal with the enormous destructive power of the atom.

As usual, Dr. Bethe led the call. "The vote against the test ban treaty undermines the entire future of arms control," he wrote. "It is a decision that should be reversed."

—W.J.B.

J. Craig Venter

The Genome's Combative Entrepreneur

Marty Katz for The New York Times

I t is like visiting a boxer in his training camp a few days before the big fight. The display of muscle. The brassy predictions of victory. The forecast that the world will be changed after the dazed opponent has been carried from the ring feet first before a jeering crowd.

But in this case the muscle consists of a HAL-sized computer, said to be the second most powerful in the world, and a ballroom-like chamber in Rockville, Maryland, that is being equipped to receive $30 million worth of high-tech machines. The rivals are the Federal Government and the world's largest medical philanthropy. And yes, the world might really be changed just a little if the vaunted victory is achieved.

Dr. J. Craig Venter, a pioneer in the sequencing, or decoding, of DNA, is beginning a new venture of boundless ambition. Sequencing the entire three billion letters of human DNA by the end of 2001, a goal first announced last May, is just the start of his grand design.

His immediate goal is to sequence the genomes of five scientifically important species—fruit fly, mouse, human, *Aridopsis* mustard and rice. These will be the core of a biomedical and an agricultural database.

His wider plan is to sequence the genomes of a thousand major species in the next 10 years to lay the foundation of an electronic information empire. An annotated genome—explanations of what each gene does, keyed to its DNA sequence—is the logical organizing principle around which to develop a full description of any organism and its biology.

The database will include DNA sequences from many sources, along with special programs for searching and analyzing the vast trove. Dr. Venter hopes it will become as essential for biologists and doctors as the Bloomberg data are for the financial world.

Trouncing the publicly financed effort to sequence the human genome seems to be just a small part, however pleasurable, of this grand design. "My major mental goal right now is not to beat publicly funded scientists at doing something," Dr. Venter says while showing a visitor around his new plant here, in a suburb a few miles north of Washington's Beltway. "What we are doing gives us far more than that."

Dr. Venter asserts that the public effort to sequence the human genome, by a group of academic centers financed by the National Institutes of Health and the Wellcome Trust of London, has chosen a flawed strategy that will produce a seriously incomplete DNA sequence. The public consortium breaks DNA into large fragments, maps the positions of the fragments on the chromosomes and then sequences the fragments. Dr. Venter's approach is to fit the fragments together without a map.

"Both Morgan and Collins are putting good money after bad," Dr. Venter says, referring to Dr. Michael Morgan, the Wellcome Trust's executive for genome sequencing, and Dr. Francis Collins, director of the NIH's part of the human genome project. "Their plan is not to finish the genome," he says, but rather to prevent him from obtaining the only sequence and locking it up with patent claims.

An associate of Dr. Collins, Dr. Kathy Hudson, said that he remained "totally confident of the current human genome project strategy" and that, "while Venter is certainly a fascinating personality, his every utterance does not seem to warrant a story in *The New York Times*." Dr. Morgan said the public consortium's strategy was safer than Dr. Venter's. So far the public consortium has an impressive record of meeting its major goals on schedule and within budget.

Dr. Venter has always reveled in portraying himself as an outsider to the biomedical establishment, a maverick implementer of bold ideas that other scientists say cannot work. Coming from almost anyone else, the rhetoric might be sharply discounted. But at least as far as his own plans are concerned, Dr. Venter has acquired both the track record and the resources—half a billion dollars—to be taken seriously.

Dr. Venter came to genome sequencing by a circuitous route. He nearly dropped out of high school in San Francisco to devote his life to surfing. He became a champion swimmer, then a Navy medical corpsman in Vietnam. That led him to enroll as a premedical student at the University of California, San Diego, but he was deflected from medicine into biochemical research.

"I've always let the results tell me where to go next," he told a reporter a few years back. "My favorite hobby is sailing. I am always tacking."

As an unknown researcher at the National Institutes of Health, Dr. Venter became interested in sequencing human DNA, particularly after the first commercial sequencing machines became available. In 1992, a medical financier, the late Wallace Steinberg, spotted the significance of Dr. Venter's ideas and set him up to work independently, founding the Institute for Genomic Research.

He and a colleague, Dr. Hamilton O. Smith, changed microbiology in 1995 when for the first time they decoded the genome of a bacterium known as *Haemophilus influenzae*. The feat sealed Dr. Venter's standing as a serious scientist. (Dr. Smith already had a Nobel Prize in physiology or medicine.) The genomes of almost every significant pathogenic organism are now being sequenced, and young scientists are flocking to what had been seen as a mature field.

Last year Dr. Venter, who is now 52, amazed the biomedical world by saying he would sequence the human genome ahead of the public consortium. Together with a colleague, Dr. Michael W. Hunkapiller, he has in effect hijacked a major company, the former Perkin-Elmer Corporation. Now the PE Corporation, it has sold off its well-known instrument-making operations and transformed itself into a pure genome company. Dr. Hunkapiller is head of one division, PE Biosystems, which makes the leading brand of DNA sequencing machines, and Dr. Venter is head of the other, known as Celera.

Dr. Hunkapiller, who first proposed the idea of a separate human project, has in effect set off an arms race between Celera and the public consortium to buy his new sequencing machines. So far Celera has ordered 300 machines and the public consortium 200. The machines cost $300,000 apiece.

Celera is capitalized at $330 million and will receive $75 million from PE Biosystems. Two pharmaceutical companies, Pharmacia and Amgen, have taken out $50 million, five-year subscriptions to Celera's human genome database. As a result Dr. Venter, who owns 5.5 percent of Celera, has more than $500 million to invest in implementing his design.

Dr. Venter calculates his 300 sequencing machines will be able to decode 140 million units of DNA every 24 hours, making Celera the most powerful sequencing center in the world. To reconstruct genomes from millions of small pieces of DNA, and store vast archives of assembled sequence, he has invested heavily in data processing equipment.

By year's end, Celera's computer will have 20 trillion bytes of disk storage. When completed, it will be the second most powerful in the world, after the Energy Department computer that simulates nuclear explosions, said Keith Pillow, a spokesman for Compaq, which supplied the computer.

At Celera's Rockville building, 50 of Dr. Hunkapiller's new sequencing machines are already running. Their first major tasks will be to sequence the genome of the *Drosophila* fruit fly and then, if this pilot project succeeds, to tackle the human genome.

Dr. Venter plans to sequence the genomes of five individuals, to be selected from the major ethnic groups and including three men and two women. They will be chosen from a pool of volunteers and will not know they have been selected.

From these 10 sets of chromosomes (each individual has a paternal and maternal set) Celera will generate a "consensus" human sequence as well as an expected 20 million or so of the DNA variations, known as polymorphisms, that make each person unique. The polymorphisms are of great interest to drug companies because they are the genetic reason why people respond differently to different drugs.

The polymorphism data will be open only for a fee but the consensus human DNA sequence, Dr. Venter said, will be made freely available for academic scientists to search, though not to download until the full sequence is completed. This restriction seems designed to prevent the public consortium from using Celera's human genome sequence to help complete its own.

The restriction creates a lack of reciprocity because Dr. Venter fully intends to use the interim sequence information being released by the public consortium. Dr. Venter said progress on the human genome would be faster if Celera and the public consortium were to cooperate, as they are doing on *Drosophila.*

Cooperation is less simple than it may appear, however, as the two groups have different rules for releasing information and different agendas. Both Dr. Morgan and Dr. Collins said they would be happy to cooperate if Celera would agree to full and immediate publication of its sequencing data.

After the human and mouse genomes are sequenced, Celera's phalanx of sequencing machines will turn to other interesting species. "I am dying to add chimp to this," Dr. Venter says. The chimpanzee genome, which differs hardly at all from human DNA, would help identify the handful of genetic changes specific to people.

"It's easy to make a list of a hundred insects whose genomes would be interesting," he says. "Look at how much money we spend on killing insects." The cow genome also seems a strong candidate, in the belief the beef industry would pay well for it.

Despite the confidence that Dr. Venter projects, Celera's approach to deciphering the human genome is not without considerable risk. Each human chromosome is a single DNA molecule roughly 100 million bases in length. The sequencing machines that determine the order of DNA bases cannot handle pieces longer than 600 bases or so. Reconstructing a full-length chromosome from such short snippets is a formidable task by whatever method it is done.

In the public consortium's approach, a chromosome is broken down into large overlapping fragments, and biologists then try to figure out which region on the chromosome each fragment comes from, a process known as mapping. In Dr. Venter's process, called a whole genome shotgun, the entire genome is broken into sets of large and small fragments, and the pieces are assembled from their overlaps without prior mapping of their chromosomal positions. Dr. Venter and his colleague Dr. Smith have developed some clever tricks to help bypass the tedious mapping stage but nonetheless will have to piece together a jigsaw puzzle of 70 million pieces without any crib sheet.

Dr. Venter's criticism of the public program is based on his view that its mapping process is inefficient. But there are risks in his own approach, too. Because the human genome contains many regions where the same DNA sequence is repeated over and over with few or no variations, some biologists believe it may be hard to reconstitute the genome from a shotgun experiment as the repetitive regions will yield a nightmare of almost identical pieces.

Dr. Venter and Dr. Smith used their shotgun strategy to sequence the first bacterial genome in 1995, but that possessed a mere 1.8 million units of DNA. The *Drosophila* fruit fly, with a genome of 160 million units, will prove a harsher test of the strategy, and even *Drosophila's* genome is a fraction the size of the three-billion-unit human genome.

Dr. Venter expresses no doubt that his enormous gamble will be successful. Sitting in Celera's cafeteria, clad casually in jeans and more relaxed than usual, he banters lightheartedly with colleagues while dishing out zingers to the opposition.

"Far from us being the evil empire, we as a corporation are providing them with the tools and reagents to try to compete with us," he says with Vader-like finality.

"Assuming we could do the *Drosophila* genome, we think the world will change," he says. "It will be like before and after *Haemophilus*. And once the paradigm is proven again, I guarantee that everybody will adopt it."

—NICHOLAS WADE
May 1999

GENOMES FOR EVERYONE

In 1990, when the National Institutes of Health began its project to sequence the human genome, Dr. J. Craig Venter was a little-known researcher working in a different section of the agency on a routine search for a heart protein that responds to adrenaline.

Just one decade later, Dr. Venter has become president of a company that seems likely to match or beat the government's effort to sequence the human genome.

This remarkable trajectory has followed a logical enough course, but one also shaped by Dr. Venter's shrewd appraisal of the technical possibilities and by his knack for forming fruitful alliances with the right individuals. It has also been influenced by the ambivalence of his relationship with the academic scientists whose approval he would probably like but whose egos he cannot help pricking from time to time.

When the venture capitalist Wallace Steinberg first offered to set him up in a company, Dr. Venter refused. Seeing himself as a member of the academic community, not a business executive, he said he preferred to run a nonprofit organization. In response, Dr. Steinberg in 1992 set up two organizations: the nonprofit Institute for Genomic Research, headed by Dr. Venter, and a company, Human Genome Sciences, headed by Dr. William Haseltine, which provided the support for Dr. Venter's institute.

The differences that soon developed between the two scientists were somewhat contained until Dr. Steinberg's death in 1995. Two years later, TIGR—Dr. Venter's playfully aggressive acronym for his organization—had raised enough funding of its own to render support from Human Genome Sciences no longer essential. So definite was Dr. Venter's dislike of working with Dr. Haseltine that he forfeited $38 million in guaranteed payments to sever TIGR's bonds with Human Genome Sciences.

TIGR's principal achievement was to rejuvenate the mature field of microbiology by showing that bacterial genomes could be sequenced. Almost every major disease organism is now being sequenced, either by TIGR or its many imitators.

But having escaped from one commercial entanglement, Dr. Venter soon plunged headfirst into another. Dr. Michael W. Hunkapiller at Applied Biosystems had developed a less labor-intensive DNA sequencing machine and reckoned it should be possible to sequence the whole human genome from scratch using the same method that Dr. Venter had pioneered for sequencing the genomes of bacteria. Recognizing Dr. Venter's management skills in building up TIGR, Dr. Hunkapiller invited him to collaborate in tackling the human genome. With the support of Perkin-Elmer (now renamed the PE Corporation) which owned Applied Biosystems, a new company called Celera was founded in May 1998.

Some 18 months later, Celera was up and running, the genome of the *Drosophila* fruit fly had been sequenced as a pilot project, and enough human DNA fragments had been generated to start the final assembly program of the human genome.

Celera has also built up a team of programmers skilled in the special art of writing programs to analyze genetic information. This new discipline, known as bioinformatics, is expected to be a key component both in understanding evolution and in searching for genes of interest to pharmaceutical companies.

Dr. Venter's grand vision is to combine Celera's computers, sequencing machines and bioinformatic services into a virtual encyclopedia of biological knowledge that everyone will want to consult.

If biology plays the same shaping role in the 21st century that the transistor did in the 20th, there's every chance that one of the pioneers of the age will be seen to be Dr. J. Craig Venter.

—N.W.

Benjamin S. Carson

Neurosurgeon and Folk Hero

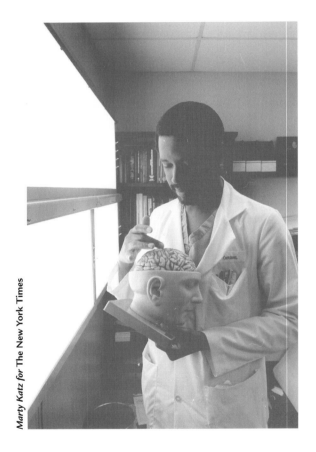

Marty Katz for The New York Times

When he first went to Johns Hopkins Hospital in Baltimore as a neurosurgery resident, Dr. Benjamin S. Carson was occasionally mistaken for an orderly. "It wasn't deliberately racist," he recalled, without a trace of bitterness. "It's just that orderlies were the only black hospital employees these people had ever seen before."

Today, at age 41, eight years after becoming the nation's youngest chief of pediatric neurosurgery—he is one of only three African-Americans in that position—Dr. Carson is arguably the most famous surgeon on the Hopkins staff. As director of the 22-member team that in 1987 successfully separated Siamese twins joined at the head, he was in the national news for weeks.

As a result, he has become something of a folk hero in black neighborhoods, no less for his own story of triumph over adversity than for the dramatic operations he has attempted.

His surgical challenges now are less bizarre than the Siamese twin separation. But they are still technically complex and emotionally demanding.

As do other neurosurgeons at leading medical centers, Dr. Carson tackles the toughest cases, from congenital dwarfism to inoperable brain tumors, from intractable epilepsy to unremitting facial pain. He tries to wield his scalpel in ways that will restore the body to something approaching normalcy.

"My philosophy is to look at a patient and ask, 'What is the worst that could happen if we do something?'" he explained during a rare free moment in his modern, uncluttered office. "It's usually that the patient ends up seriously debilitated or dead. Then I ask, 'What is the worst that could happen if we do nothing?' And it's usually the same thing. So with that as a background, I figure it's always worth trying to do something, if there's any chance at all that doing something might end up helping."

This attitude led Dr. Carson, in 1985, to revive the surgical procedure known as hemispherectomy, the removal of half the brain in a child who is plagued by seizures that do not respond to drugs. The hemispherectomy, first developed in the 1930's, is an operation of epic consequence, performed in the hope that the remaining hemisphere can orchestrate thought, speech and movement for the whole body.

After several hundred unsuccessful attempts, the procedure fell into disfavor in the 1970's, not because of the dysfunction resulting from the loss of brain tissue, but because of the nearly inevitable post-surgical complications, including bleeding, infection and problems with the cavity left behind.

But when Dr. Carson met Maranda Francisco, a 4-year-old girl from Denver who was racked by 120 seizures a day, he decided that medicine had advanced sufficiently to give the measure another try.

"The chief of pediatric neurology here, John Freeman, felt that Maranda was a perfect candidate for a hemispherectomy, which he had seen performed a few times while he was at Stanford," Dr. Carson said. "So I did a lot of reading on the subject, and it seemed to me the complications were mostly things we could handle now."

The odds against success were great, but Dr. Carson figured it was worth taking a chance. "I reasoned that she was having so many seizures that she had no life, so there was not really anything to risk," he said. "And there might be a whole world to gain."

Maranda is now a healthy 12-year-old who takes tap-dancing lessons. Like most of the other hemispherectomy patients at Hopkins—44 in all, which the hospital believes is more than at any other center in the country—Maranda got her speech back immediately after surgery, probably because her right hemisphere had already taken over language function from the badly damaged left hemisphere.

The main physical consequence of the loss of half a brain—paralysis of the opposite side of the body—was corrected in Maranda's case by several months of physical therapy. Most of the other children have had similar restoration of speech and movement. Dr. Carson estimated that 80 percent of the hemispherectomies done at Hopkins have significantly reduced or eliminated the patients' seizures.

"We can only do this operation on young children, because their brain cells haven't decided what they want to be when they grow up," Dr. Carson said. This flexibility, which neuroscientists call plasticity, explains why certain brain cells take over the functions of damaged or missing cells.

Heroic as these hemispherectomies have been, the real media attention came after the Siamese twins case, which held a sort of lurid fascination for physicians and laypeople alike. When the parents of the twins, Patrick and Benjamin Binder of Ulm, Germany, sent in a request to have their sons operated on at Hopkins, there had never been a successful separation of Siamese twins joined at the head.

It was just the kind of long shot on which Dr. Carson seems to thrive.

The operation, for which the surgical team spent five months rehearsing, lasted 22 hours and involved 70 surgeons, nurses and assistants inside the operating room, and another 70 support staff members outside.

During the operation, Dr. Carson used a new way of buying time for brain repair. It involved hypothermia, cooling the patient's body to 68 degrees to slow brain metabolism sufficiently to allow the surgeons to stop blood flow through the brain for an hour without causing damage.

"I almost didn't make it," he recalled. "After they stopped the heart and set the timer for one hour, I told everyone I didn't want to know how much time we had left."

He managed to separate the two brains in about 20 minutes and to close the skull of the twin he was working on—with his mentor, neurosurgery chief Dr. Donlin Long, closing the other twin's skull—in about another 40.

After the operation, the successfully separated twins went home to Ulm. But, German doctors said, the boys were too disabled to live at home and were placed at a facility for handicapped children in the area.

"I still believe the surgical procedure itself was an incredible success," Dr. Carson said. "And whenever there's a less-than-ideal result, I feel you must always ask yourself, 'What can I learn from this? How can I do better next time?'"

Although Dr. Carson has occasionally been criticized for publishing his research only infrequently and has had his originality as a scientist questioned, he is unruffled by such comments. A mild, soft-spoken man with an unhurried manner and a ready smile, he projects confidence, which he attributes to knowing his subject cold, believing in his talent and relying on his religious faith.

A Seventh-Day Adventist, Dr. Carson said one of his favorite sayings is, "Do your best, and let God do the rest."

The most dramatic procedures he performs today involve large brain tumors. When the size of these encroaching tumors is significantly reduced, Dr. Carson said, the stray pieces of cancer that are impossible to remove surgically are often destroyed, even without dangerous radiation therapy.

"If you can reduce the bulk of the tumor and shave it back to normal tissue," he said, "then I believe the body's natural defense mechanisms can keep the rest of it under control."

Beating back nearly insuperable odds, which is what he attempts in the operating room, is also something of a theme in Ben Carson's personal story. His father abandoned the family when Ben was 8, and he grew up in a Detroit ghetto, where his mother provided for her two sons by working several domestic jobs. After a rocky start in elementary school—he laughs when he recalls that he was, by universal consensus, the "class dummy" in fifth grade—he began to excel academically, a fact that he attributes to his mother.

"She made my brother and me turn off the television, restricting us to two or three programs a week," Dr. Carson said. "And she made us read two books a week from the Detroit Public Library, and write book reports for her." It wasn't until years later that the boys discovered that their mother, with a third-grade education, could not read those reports.

This demanding regimen soon made a mark on young Ben Carson, who graduated third in his high-school class and won an academic scholarship to Yale University, where he majored in psychology.

Now, in between operations and hospital rounds, Dr. Carson, the father of boys whose ages are 9, 8 and 6, frequently delivers inspirational lectures. And he sets aside an hour every month to talk to schoolchildren.

At the most recent such lecture, 750 youngsters from nine schools greeted Dr. Carson's arrival in the Hopkins auditorium with the fanfare usually reserved for rock stars: squeals, applause and requests for his autograph or a chance to snap his photograph.

"You're my role model," one teenager told him. "I want to be a brain surgeon, too."

"You hold on to that dream," replied Dr. Carson. Still trim and with a full head of hair, he appears so youthful that in his blue surgical scrubs and white coat he almost looks like someone masquerading as a doctor.

But no one mistakes him for an orderly anymore.

At the lectern, his message was simple: "Don't let anyone turn you into a slave. You're a slave if you let the media tell you that sports and entertainment are more important than developing your brain."

The hands that have made him one of the most sought-after neurosurgeons in the country, the hands with the long tapering fingers and immaculate oval nails, gripped the microphone tighter as he delivered the message that has become almost as important to him as the operations he performs.

"You don't have to be a brain surgeon to be a valuable person," Dr. Carson said. "You become valuable because of the knowledge that you have. And that doesn't mean you won't fail sometimes. The important thing is to keep trying."

—ROBIN MARANTZ HENIG
June 1993

LOOKING FOR MORE BEN CARSONS

In the late 1980's, when Ben Carson was drawing acclaim for his skill in the operating room, his career was extraordinary not just for his achievements but because it was a black man who was making them. No one knew better than he just how much of a rarity he was in the elite circle where surgeons dwell, and Dr. Carson, the product of a Detroit ghetto, made a point of taking time to lecture young people in the hope of changing the face behind the surgical mask.

Not much has changed since then. Dr. Carson is still a much sought-after surgeon and speaker. And he is still largely alone. Indeed, some studies suggest that the number of African-Americans who become surgeons has actually been dwindling—a loss of ground that dismays those in medicine who had thought things could only get better.

"Diversity is not a real happy word among surgeons," says Dr. William Lynn Weaver, a professor of surgery at Morehouse University, the historically black institution in Atlanta. "What I've been trying to do is get the American College of Surgeons to admit that there is a problem."

No one can argue with the numbers. The most recent survey by the American Medical Association showed that in 1998, only 408 of the 5,065 surgical residents in the United States were African-American. That was a slight improvement from the year before, but other studies show that, in general, fewer black students are trying to become surgeons. Indeed, the National Medical Association, a group formed in 1895 to fill the needs of black doctors and patients, has officially shelved its goal of having "3,000 by 2000"—that is, having 3,000 minority students in American medical schools by the turn of the century. And the last survey showed that there were zero black students going into pediatric surgery, Dr. Carson's specialty.

The successful attack on affirmative action programs, Dr. Weaver and others said, virtually erased in two years what had taken two decades to accomplish. At the same time, black students who seem like good candidates for becoming surgeons are also being enticed by other areas of medicine, especially primary care, and by other industries altogether, like high technology. In the face of this competition, some black surgeons say, the surgery establishment has simply failed to be aggressive enough in courting prospective applicants. It does not help that, at last count, of the 125 medical schools across the country, only 6 had surgery chairmen who were black (and 4 of those were minority medical schools). Indeed, some found demoralizing a 1999 editorial in a journal of surgery that discussed the lack of black surgeons and declared that however desirable it might be to have more black surgeons, "the American College of Surgeons has not adopted a population parity standard as a goal." (This position also dismays some people who note recent medical studies suggesting that minority doctors are often more sensitive to their patients' needs than are white doctors.)

One step toward bridging the gap is mentoring programs in which black surgeons encourage medical students to join the profession, said Dr. Lenworth M. Jacobs, a Connecticut surgeon who is president of the NMA's surgical section. But he said the encouragement should begin even earlier. "The way you achieve that is by going into the high schools and providing role models and opportunity for high school students," he said. "You can't just start at the end and go forward."

Still, in the absence of a clear effort by the overall community of surgeons to diversify its ranks, some experts remain skeptical that much progress will be made soon. And that, said Dr. Weaver, is a loss not just for African-Americans but for medicine. "There are other Ben Carsons out there," he said.

—Eric Nagourney

Abhay Ashtekar

Taste-Testing a Recipe for the Cosmos

Michael Tweed for The New York Times

Dr. Abhay Ashtekar, the leader of a worldwide effort to unify the two most profound, abstract and mathematically baroque theories of physics discovered in this century, is sprinkling frozen mango cubes on scoops of vanilla ice cream. After a meal of fish and tamales during which he puttered about his kitchen in a green apron emblazoned with bright flowers and the words "golden poppies," Dr. Ashtekar serves the dessert to his wife, Christine Clarke, and two visitors, while needlessly apologizing for his cooking skills. "I have never been able to follow a recipe," he says ruefully. "I just add what seems right."

The irony in this bit of self-criticism is lost on no one at the sun-splashed kitchen table. A flair for working without a recipe is a necessity for pursuing the central scientific passion of Dr. Ashtekar, who heads the Center for Gravitational Physics and Geometry at Pennsylvania State University and is temporarily living in Santa Barbara, California, to help organize a six-month workshop.

His passion boils down to this: an attempt at creating a sort of cosmic nouvelle cuisine by merging Albert Einstein's theory of general relativity with the laws of quantum mechanics, which were first worked out in the 1920's by a number of physicists including Erwin Schrödinger, Paul Dirac, Werner Heisenberg, Niels Bohr and Einstein. No recipe exists, and only a few of the ingredients are known.

Between them, these theories seem to explain the observed universe, but they express profoundly different conceptions of matter, time and space. That philosophical schism also leaves physicists in deep doubt over how to deal with phenomena in which both theories should be valid—in the realms of the very small and the very energetic, like the Big Bang in which the universe was allegedly born.

"We have two wildly successful theories that have defined 20th century physics," said Dr. Gary Horowitz, a physicist at the University of California at Santa Barbara. "These theories are fundamentally incomplete and inconsistent with each other, and we just can't go on like that."

Relativity theory describes how the gravity of everything from subatomic particles to massive stars distorts and curves the four dimensions of space-time, like coconuts rolling on a rubber sheet. That changing curvature, in turn, determines exactly how the objects orbit about one another or fall together. A large enough congregation of matter can collapse to a point of infinite density, called a singularity, and shroud itself in a sphere of darkness—a black hole, whose gravity is so powerful that nothing can escape from it, not even light.

On the other hand, standard quantum mechanics tells the tale of a "flat" space in which particles refuse to orbit smoothly; instead, they can hop suddenly from one spot to another, carrying with them only specific, sharply defined amounts of energy called quanta, like tourists holding no bills smaller than a 20. And far from respecting the crisp determinism of classical relativity, these particles sometimes exist not at definite positions but rather as fuzzy clouds of probability.

In culinary terms, these two kinds of physics have remained as distinct as a Tex-Mex barbecue and a New Age vegetarian picnic taking place in the same park. But most physicists, like Dr. Ashtekar, believe that since there is just one universe, there should be just one fundamental way of describing it.

Dr. Roger Penrose of the University of Oxford said that both theories also had internal problems, like the strange singularities that form in general relativity and the unmanageable infinities that also crop up in quantum mechanics. "The expectation is that to resolve these issues, we need the correct union between general relativity and quantum mechanics," Dr. Penrose said. "In my view," he added, Dr. Ashtekar's approach "is the most important of all the attempts at 'quantizing' general relativity."

That approach has led to a daring conception of space-time that shares characteristics with both the quantum world and general relativity. On incredibly tiny scales—10^{-33} centimeters, or smaller than a trillionth of a trillionth of the diameter of an atom—space-time becomes jagged and discontinuous. At those scales, Dr. Ashtekar said, space dissolves into a sort of polymer network, "like your shirt," which looks continuous from a distance but is actually made of one-dimensional threads.

These developments had their start in the 1980's with a mathematical reformulation of Einstein's theory by Dr. Ashtekar, which allowed it to be molded into something that looked like a modified quantum theory. The equations that resulted were later shown to predict the polymers by Dr. Ashtekar, Dr. Carlo Rovelli of the Center for Theoretical Physics at the University of Marseilles in France and Dr. Lee Smolin of Pennsylvania State University.

Dr. Ashtekar freely points out that no one yet knows if this conception is right, whether nature has actually chosen to fricassee space at the smallest scales. His approach to unification is not even the most popular one; many elementary particle physicists, including Dr. Horowitz, the Santa Barbara physicist, think that vibrations of 10-dimensional entities called strings might hold the key to all the forces of nature, including gravity.

But because it uses relativity theory as its jumping-off point, the approach dreamed up by Dr. Ashtekar and his colleagues is "the most in the spirit of Einstein," said Dr. Thomas Thiemann, a physicist at the Albert Einstein Institute, a part of the Max Planck Society in Potsdam, Germany.

Though known as a researcher of intimidating mathematical prowess, Dr. Ashtekar, 49, seems to approach his métier without much solemnity about its cosmic implications. After lunch, he sat outside in brilliant sunlight near an orange tree, sipping tea, shuffling through visual depictions of what the fabric of space might be. On his mug, a Gary Larson cartoon showed a schoolboy raising his hand in class and asking: "Mr. Osborne, may I be excused? My brain is full."

Dr. Ashtekar's light touch is more than a matter of style. Later, when pressed, he shrugged off the almost religious significance that some cosmologists have suggested a final theory will have. "That arrogance is somewhat misplaced," he said. "I personally feel this is a great intellectual challenge; it's very difficult and all that." But physics, he said, "is only a part of the whole mystery of nature, of existence and ourselves."

Abhay Ashtekar grew up in small cities in the Indian state of Maharastra, which contains Bombay. At 15, while living in Kolhapur, a town surrounded by lush green sugarcane fields, he came across the popular book *One, Two, Three . . . Infinity* (last printed in 1988 by Dover) by the physicist George Gamow and decided that he liked mathematics and cosmology. Two years later (after Ashtekar's first year of college in India), a math professor explained to him that it was actually possible to make a living doing research on such topics.

"Coming from a middle-class family, you either became a doctor or an engineer or you entered civil service," Dr. Ashtekar said. The notion of a career fiddling with ideas "was a total revelation," he said. "And my mind was set. I wanted to try to do pure research."

He began showing a flair for not following scientific recipes almost immediately, discovering a small but significant error in the answer to a problem given in *The Feynman Lectures on Physics,* a set of introductory volumes written by the late Dr. Richard P. Feynman, a Nobel Prize winner. Dr. Ashtekar wrote to Feynman—and still has the treasured reply conceding that the textbook was wrong.

Sometime later he walked into the United States consulate in Bombay and searched university catalogues for graduate programs in gravitation. He eventually landed, a very uncertain 20 years old, at the University of Texas, Austin. On his first day on campus, he had to work up his courage just to enter the physics building. Eventually he entered, climbed a set of stairs, looked both ways down a corridor and hurried out again.

His confidence quickly returned. He went on to complete his Ph.D. at the University of Chicago, and then to appointments in Oxford, Paris and Syracuse, New York, before settling at Penn State.

Dr. Ashtekar gradually became interested in a field that many theorists before him had entered at their peril. "The history of quantum gravity," wrote Dr. Rovelli of Marseilles in a recent review, "is a sequence of moments of great excitement followed by great disappointments."

In an early attempt, Schrödinger announced in 1947 that he had managed to unify Einstein's equations with the theory of electromagnetic fields. But Einstein dismissed the work and condemned the worldwide news coverage it had attracted.

From such episodes, wrote Einstein, "the reader gets the impression that every five minutes there is a revolution in science, somewhat like the coups d'état in some of the smaller unstable republics."

Later, Feynman and others tried to slip gravity into the quantum version of electromagnetism that had been developed, but the hybrid theory exploded with mathematical infinities when all the interactions it permitted were added up. And in the heyday of a unified theory called supergravity, Dr. Stephen W. Hawking, the physicist at Cambridge University, gave a talk titled "Is the End of Physics in Sight?"

It was not, yet, and when infinities reared their head in supergravity, some of its ideas were salvaged and woven into string theory.

Dr. Ashtekar's approach, which drew in part on work in the 1960's by Dr. John Wheeler of Princeton University, began with Einstein's equations directly. Following his mathematical taste buds rather than accepted formalisms, Dr. Ashtekar searched for some way to transmute the theory's geometric spirit into the fuzzy quantum world.

His first step was to express Einstein's equations in terms of variables with a chirality, or "handedness," in which a circle drawn in the clockwise sense would look different from one drawn in the opposite sense. Against all intuition, Einstein's equations, which show no preference for direction, broke into simpler pieces in the Ashtekar variables. From that vantage, first achieved in 1986, Dr. Ashtekar was able to use straightforward tricks developed in the 1920's for creating a quantum theory from a deterministic one.

"It's rather curious," said Dr. Chris Isham, a theorist at Imperial College in London. "At one level, it was simply a redefinition of variables, which one might think was a fairly minor thing to do. But it really rejuvenated the whole field and caused quite an explosion of activity."

Soon there followed the solutions showing the polymerlike structure of space. The interlocking polymers "quantize" space in a particularly odd way, since each strand is somehow loaded with a definite amount of cross-sectional area. So to figure out the area inside a circle in this weird space, one would count up the strands that puncture its surface and multiply by the quantum of area carried by each of them. In this way, area is not smooth but comes in bundles.

The same rule holds for the area of a black hole's event horizon, the place beyond which anything is drawn into the black hole. In 1998 Dr. Ashtekar, Dr. John Baez of the University of California at Riverside, Dr. Kirill Krasnov of Penn State and Dr. Alejandro Corichi of the Universidad Nacional Autónoma de México showed that the polymers running into a black hole in a sense hold it "still" at the puncture points, like a water balloon supported on blunt pins. The rest of the horizon is free to jiggle about quantum-mechanically.

Like the bouncing and jiggling of atoms in an ordinary gas, such motion has a definite entropy (or randomness) and therefore a temperature. Drawing on earlier formulas obtained with Dr. Jerzy Lewandowski of the University of Warsaw, Dr. Ashtekar and his colleagues were able to calculate the entropy for a black hole, matching a famous 1974 prediction by Dr. Hawking. The theory has other bizarre consequences, such as a bending of space that could be caused by immensely energetic photons of light. But the theory still faces challenges, particularly in the treatment of the time part of space-time. Until better observational tests turn up, the furious bake-off whose prize is unifying physics will remain a theoretical one. The entrants are not limited to strings and quantum geometry, which are beginning to look almost conservative. Dr. Penrose of Oxford believes that quantum mechanics itself will have to be modified before a fully successful merger with relativity can be made. Dr. Rafael Sorkin of Syracuse University is starting from scratch, postulating bits of existence he calls "the atoms of space-time" and working upward toward the macroscopic laws of physics.

In his side yard, Dr. Ashtekar, after a brief disappearance, brings a late-afternoon snack for his visitors. It is a reminder that no matter what package of formulas emerges, he is likely to be kneading and rolling them in his mathematical kitchen, so to speak, with results that could not be predicted by looking at the picture on the box.

—JAMES GLANZ
April 1999

THE RETURN OF GRAVITY

For most of the past hundred years, the study of gravitation has been the siren song of physics. Like Ulysses, who ordered his crew to stop their ears with wax and lash him to a mast so that he could hear the Sirens' song and survive, only Einstein seemed able to submit to gravity's charms and find success in his research.

Gravity is so much weaker and so different from the three other known forces of nature that attempting to unify it with the rest of physics seems, in retrospect, to court failure. Most ominously, even Einstein made little progress after 1915, when he formulated the theory of general relativity, which describes how gravity bends space on large scales. He spent fruitless decades trying to merge gravity and electromagnetism—just one of those other forces—into a single theoretical framework.

But, as the 21st century begins, gravity has become one of the liveliest subdisciplines of physics. Theorists are bringing together research on the fabric of space on the smallest imaginable scales, the geometrical shape of the entire universe and oddities that depend on especially powerful gravity, like black holes and neutron stars. Not only is understanding gravity's relationship with other forces considered crucial to the field's future; to hear physicists tell the story, they are actually making modest progress.

Gravity owes its transformation to a series of isolated and sometimes contentious successes scattered over the better part of 70 years.

By most accounts, gravity's redemption began at sea, when a young Indian student named Subrahmanyan Chandrasekhar was traveling by ship from Bombay to Cambridge, England, in the summer of 1930. He calculated that as massive stars age and cool, the force of their own gravity could cause them to collapse abruptly rather than quietly fade away.

The collapse would concentrate the gravitational fields still further and create superdense objects that astrophysicists had not even begun to think about yet. Those consequences were so extreme that Sir Arthur Eddington, the Cambridge astrophysicist who was one of the leading scientific figures of his day, publicly ridiculed the calculation and implied that it must be flawed.

But other physicists gradually made the case that those collapsed objects were probably out there in space somewhere. First, in the 1930's, the astronomers Walter Baade and Fritz Zwicky proposed the existence of the stellar cinders called neutron stars, objects in which a mass greater than the Sun's would be squeezed into a sphere about the size of Chicago.

By the 1960's, theorists had gotten comfortable with the idea that matter could collapse even further—all the way down to a point, called a singularity. Einstein's theory of relativity predicted that a singularity would bend space so drastically that nothing could escape its vicinity, not even light.

Even so, during a cross-country drive in 1966, a young theorist, Kip Thorne, still found it prudent to pay a visit Dr. Chandrasekhar at the University of Chicago, where he had settled after his battles with Sir Eddington. Were those strange and unlikely objects, called black holes, worth devoting a good fraction of one's career to?

Dr. Chandrasekhar replied that research on black holes was at least as likely to bear fruit as his original calculation of stellar collapse was. The advice seemed hard to ignore when astronomers discovered neutron stars in space the following year. Gravity's siren song was losing some of its peril.

Things picked up even more a decade later, when Dr. Stephen W. Hawking and others showed that the powerful gravity of black holes could in effect tear particles out of apparently empty space and radiate them away, as an ordinary object radiates heat. Later calculations by many physicists suggested that this intersection between powerful gravity and microscopic particles holds deep clues to how gravity should be merged with the rest of physics.

The rate of that radiation, for example, turns out to depend on just how space is chopped up on the smallest possible scales, a kind of graininess in existence that can't be completely understood until gravity is.

That radiation is so weak that it probably won't be seen directly for a long time. But Dr. Thorne has become a driving force behind the Laser Interferometer Gravitational-Wave Observatory, or LIGO, a $350 million device that aims to detect larger gravitational ripples in the fabric of space for the first time. LIGO's first observations are scheduled for 2000, and the most likely sources for those ripples are cataclysmic mergers of neutron stars and black holes somewhere deep in the cosmos.

—J.G.

Geerat Vermeij

Getting the Feel of a Long-Ago Arms Race

Running his fingers across a shell, whether the rocky spire of a 400-million-year-old fossil or the glassy dome of a modern-day cowrie, Dr. Geerat Vermeij is quietly reading tales from the history of life. Examining the shells' punctured armor, he sees every detail of their encounters with predators and their close scrapes with death or final agony. Yet at the same time Dr. Vermeij sees nothing at all, because he is blind.

Born with glaucoma and never able to make out more than fuzzy shapes, Dr. Vermeij (pronounced Ver-MAY), a paleontologist whose colleagues call him Gary, has been completely blind since the age of 3 and has never really seen a single creature, living or fossil. Yet by using his fingers to feel both the damage on shells as well as the girth and power of the claws and jaws that attack them, this professor at the University of California at Davis has found evidence of an ancient arms race. According to the histories recorded on these broken and mended fortresses, mollusks appear to have evolved ever more rugged armor to protect their delicate flesh just as their predators developed more vicious weaponry.

"Gary is a brilliant guy, an idea man, a synthesizer," said Dr. David Jablonski, a leading paleontologist at the University of Chicago. "It's very easy to think about how a predatory snail will catch a clam and kill it. But how does that play out over millions of years of ecological time? Gary is the guy who has really dug into that. His observations have swept the field and will still be cited 100 years from now. We should all be so lucky."

Researchers say paleontologists have typically ignored ecological interactions like predation, many focusing instead on how large-scale, physical factors like climate change shape life in the fossil record. Dr. Vermeij's views have forced them to rethink the importance of animals in shaping each other's evolutionary fates. Researchers say Dr. Vermeij's findings are among the foundations of the emerging field of paleoecology.

"It's anything but the romantic idea that nature is nice and kind and stable," said Dr. Vermeij. "To some people it isn't a pretty view of the world. It's nasty and things get nastier and nastier. Everyone is affected mostly by their enemies. That is indeed my view."

When meeting Dr. Vermeij, one is struck by an air of spareness. Born in the Netherlands and raised in New Jersey, this thin, almost gaunt 48-year-old man holds a visitor's attention with his quiet voice and a direct if unseeing gaze. His office likewise has a spartan feel, filled only with papers and boxes of shells, leaving the visitor to wonder how it is that this man who cannot see can manage to be an evolutionary biologist; a teacher; the editor of *Evolution,* the field's foremost journal; a MacArthur Fellow; an obsessive shell collector; a world-traveled explorer; and a field naturalist.

But within these walls, sight is entirely superfluous. Data can be taken by touch. Voluminous books can be tapped out in sheaves of Braille pages. Words meant for the sighted can be written on a typewriter and the voice of a person reading can allow the perusal of anything ever written.

But it is the observation and exploration of the living world that would appear to be the most difficult hurdle for Dr. Vermeij. Instead, researchers say it is the greatest strength of this keen natural historian, who has worked in such places as Guam, Africa, New Zealand and Panama and who at times has strayed far from his beloved shells to publish on such diverse topics as leaf shape and the evolution of birds.

In fact, he is famous for the invention of what amounts to a unique method for observing the natural world. When biologists disembark on new shores, it is largely their eyes that inform them. Life is what can be seen. But for Dr. Vermeij, life is what can be grasped, with hand or foot, and examined in every other way.

"I listen and smell and feel," said Dr. Vermeij, a man who would seem to like nothing better than for you to forget that he is blind and who strikes a triumphant note when recounting tales of exploring snake-filled swamps and wading neck-deep in oceans swimming with sharks and stingrays. "I've stuck my hands into more holes and under more rocks than I care to mention, but my feeling is if you want to experience nature you've got to be unconstrained."

Researchers describe his ability to see with his hands as "phenomenal" and say that it is what makes Dr. Vermeij's view of evolution as convincing as it is unique. Colleagues recount in awestruck tones his ability to feel differences among shells, quickly identifying them down to the level of subspecies. They tell tales of his exploring new habitats more thoroughly and with more insight than most seeing biologists.

Dr. Warren Allmon, director of the Paleontological Research Institution in Ithaca, New York, described the experience of escorting Dr. Vermeij to visit a fossil bed in Florida that Dr. Allmon had previously spent countless hours studying.

"He stood there for maybe 45 minutes feeling it," said Dr. Allmon, "and then proceeded to tell me almost everything there was to know about it— the fighting conchs, the percent covered with barnacles, a layer of oysters. Now if I took you there and said, 'Look right here. See the layer?' you still wouldn't see it. I was trying not to show my amazement since I had heard this about him, but I was really blown away by that. He can do things with his hands that most of us can't do with our eyes."

Dr. Stephen Jay Gould, a paleontologist at Harvard University, describes Dr. Vermeij as "an acute thinker and an excellent observer." He said in an interview: "He has an uncanny way of perceiving correlations between shell forms and environment. He's noticed things that other molluscan biologists never did, even the most minor changes in the shapes of shells."

It is just such perceptions, combined with what researchers describe as an encyclopedic knowledge of the world's mollusks and the literature on them, that has allowed Dr. Vermeij to pull together his wide-ranging theory on the perpetual cat-and-mouse games that he sees as steering the course of evolution.

Dr. Vermeij's view is based not on a single breakthrough experiment. Rather it was slowly formulated over two decades as he, often working with his wife and colleague, Dr. Edith Zipser, also at the University of California at Davis, traveled the globe in search of the histories written on broken shells.

A big piece of the puzzle was live-to-tell scars, repaired breaks where predators had tried but failed to get in. Looking at shells as old as 300 million years, Dr. Vermeij and his colleagues found that ancient shells showed fewer repairs and recent shells showed more. Seeing these permanent records of narrow escapes growing more frequent over time, the researchers suggested that the ability to escape predators was becoming an increasingly important key to survival. Dr. Vermeij also noticed that those shells whose geometry made them more vulnerable to breakage seemed to be on the decline or restricted to places like the deep sea where modern, well-armed predators did not abound.

Examining risky habitats full of light and life and predators, he documented evidence of architectural innovations in more modern shells, changes that helped to protect a shell's inhabitant. Over time shells appeared more tightly coiled, with greater buttressing, showing more teeth lining increasingly smaller openings in heavily guarded entryways.

Meanwhile, the fossil record showed crabs, fish and others who would dine on these shelled delicacies diversifying and becoming better at cracking, popping, drilling and peeling their victims open—all mounting evidence in favor of a hypothesis that predicts that life on Earth will only get more and more difficult.

Throughout his career, Dr. Vermeij has had to cope with opposition from people who doubted that a blind shell enthusiast could become a scientist. The list runs from government officials who initially refused to give him scholarship money to hire readers so he could study biology at Princeton, to the faculty at the University of Maryland at College Park who gave him a trial position before allowing him to assume a tenure-track job. In an interview for graduate school at Yale University, a biology professor made Dr. Vermeij submit to a practical test.

"He asked me how I could possibly read the literature, how I could possibly do this, that and the other thing," said Dr. Vermeij. "He decided that the clincher would be for me to fail at the thing I was supposed to be the best at, shells. So he took me down to the Yale Peabody Museum collection. He pulled out a couple of shells and asked me what they were and I got them right, fortunately. I am forever grateful to someone flexible who, like any good scientist, can change his or her mind."

With Dr. Vermeij firmly established as a scholar, researchers now say that it may be his blindness and the tactile talents which he developed because of it that have allowed him to see what others do not.

Even Dr. Vermeij himself acknowledged that his view of the living world must ultimately be different. "It is interesting and I really haven't come to grips with it," he said. "I know that when I look at shells I look at them tactilely. I know that I see characteristics differently from other people. For example, when I read people's descriptions I think they're just not describing what I see and in turn I see things that they don't. Feeling shells does predispose one to noticing some things that other people would be less aware of, so that has played a role."

After a short time talking to a person who has accomplished all he set out to do without the aid of sight, one soon begins to come to Dr. Vermeij's opinion that blindness, in itself, is a rather dull point of discussion. Self-described as a "lifelong opponent of affirmative action" for all minorities, he takes as harshly unsentimental a view of the blind as he does of the evolution of life.

"I am a strong disbeliever in seeing things from the point of view of being handicapped, gender, race and all the rest of it," said Dr. Vermeij, adding that he hurried through Princeton and Yale in three years each on full scholarships so as not to be a "parasite."

"All my life I have fought hard to integrate into society and I think that's the way any minority group should work," he said. "If you give people preferential treatment, others will always say, 'Ah well, he got this because he's blind.' You can never live that down. The idea is to eliminate the barriers to the point that nobody will care."

And perhaps, in his case, that point has been reached.

"With anyone who's apart any way from the norm you notice it at first," said Dr. Gould. "But after a while, he's just Gary."

—CAROL KAESUK YOON
February 1995

MARRYING EVOLUTION AND ECONOMICS, GINGERLY

Most scientists proceed through their careers becoming increasingly specialized, ever more restricted to the tiny fascinations of a particular organism or pet theory. But Dr. Geerat Vermeij, who was already looking at the big picture by studying hundreds of millions of years of evolution, is now stepping back to take a still broader view in what he says could be the most significant contribution of his career. Seeking answers to questions about the history of life, he has begun looking beyond biology, beyond even the natural sciences to the remote reaches of the world of economics.

Convinced from his earlier work that the principles underlying the evolution of life and the development of economic systems are one and the same, Dr. Vermeij—who has been working on this project for years but describes his efforts as still preliminary and experimental—is hoping that a study of economics will help illuminate not only the history of life, but its future as well.

The idea is that economies, entrepreneurs, whole businesses and countries act and react like living organisms. Like living organisms, these economic entities use resources. Rich and powerful corporations dictate the way much of the world's business is conducted, how money flows and what people have available to buy. In the same way, key dominant species, known as keystone species, dictate much of what goes on in a given ecosystem, how the key resources—nutrients and energy—flow and what other organisms are present and how they fare.

In the same way, ecosystems are dependent on one another. For example, some coastal ecosystems are entirely dependent on the energy and nutrients imported into them from ocean ecosystems. As sea birds gather fish and other rich resources of food from the ocean, they deposit resources back on the beach in the form of feces and sometimes their own bodies as well. If the ocean ecosystem crashes, it will take the coastal ecosystem right along with it, just as some economies are dragged down when those they depend on fail.

"Although consumers are supposed to have free choice, what they buy is largely determined by advertisers who have more power and information at their command," he said. "That's not so different from predators that determine much about the lives of their prey."

It is an obvious idea that, while far from new, has been only little and poorly explored, Dr. Vermeij said.

"There are several books with economics and evolution in the title, but they're pretty bad," he said of treatises written by economists dabbling in evolution. To correct this, Dr. Vermeij has been reading everything he can find on economics. "Just as I'm a novice in economics, they're novices in evolution and they're thrashing around in the dark. I recognize the danger here to make a fool of myself."

But while there are risks, including what Dr. Vermeij says is the possibility that nothing may come of this exploration at all, the rewards are potentially huge. In addition to trying to understand patterns in the evolution of life better by drawing lessons from economics, Dr. Vermeij hopes to learn more about the future of the world's economies and the biosphere.

One of the things that the biosphere and the world's economies have in common is that they are characterized by continued, expansive growth. With a few setbacks here and there, like mass extinctions that temporarily check the rise in species numbers, the diversity and abundance of species on Earth have continued to grow ever since life began. Likewise, the world economy has continued to grow. But given the limited resources that ultimately feed both these systems, many say such unbridled growth simply cannot continue forever.

The problem is that when growth slows and stability sets in, opportunities decline, whether for organisms trying to exploit new ways of living in an ecosystem no longer flush with resources or for people seeking to start new businesses or create jobs in an economy whose opportunities are becoming fewer and further between.

"The interesting question," said Dr. Vermeij, "and one for which I don't have a good answer yet, is if we can't grow forever, how can we break the rules to break that awful historical precedent?"

Coming up with entirely new ways to steer world economies and the biosphere away from decay and keep them forever on a course of rich opportunity and innovation is a tall order, to say the least. But Dr. Vermeij may well be on the right track, if his methods are any indication. Another reasonably successful evolutionary biologist drew on ideas of the economist Malthus to come up with his theories, that biologist being Charles Darwin himself.

—C.K.Y.

John H. Conway

At Home in the Elusive World of Mathematics

Dith Pran/The New York Times

D r. John H. Conway sits down at his computer and gets ready to log on. But before the computer allows him to begin work, it quickly spews out 10 randomly selected dates from the past and the future, dates like 3/15/2005 or 4/29/1803. Dr. Conway has to mentally calculate what day of the week each would be before his computer lets him open a file and get to work.

It is a game he has rigged up to play with himself. "I think I'm the fastest person in the world at this," he says. His record is 15.92 seconds to calculate all 10 days. But, he modestly says, it is not that hard. "It's the kind of thing autistic kids do."

Dr. Conway is unlike any other mathematician at Princeton University, and even in the field at large he stands out for the originality of his work on a mix of disparate and difficult subjects, including number theory, knot theory, the theory of quadratic forms and the theory of games.

And in a profession where quirky skills, like being a lightning calculator, are not unexpected, he is also remarkable for his odd and unusual interests and abilities. He is obsessed with games and puzzles, and has a thirst to learn unusual facts, like the digits of pi extended to 1,000 places or the name and location of every star in the heavens. "It took me a year to learn that," the 55-year-old mathematician said, noting that the sky changes each night. But at the same time, he has made significant mathematical discoveries of both theoretical and practical importance.

Dr. Ronald L. Graham, adjunct director for the information science division at AT&T Bell Laboratories in Murray Hill, New Jersey, described Dr. Conway as "one of the most original mathematicians" around. "He's definitely world class, yet he has this kind of childlike enthusiasm," Dr. Graham said. "He's confident enough to work on any crazy thing he wants to." He added that Dr. Conway was entitled to veer off the beaten path. "He's certainly achieved enough," Dr. Graham said.

But Dr. Conway has so many peculiar interests and quirks, and he can so easily be made to sound eccentric, that his deep love of mathematics and the natural world can be lost in the fluff. In a way, he is oblivious to the routines and customs of ordinary life. He recently bought a pair of shoes, after wearing only sandals, even in winter, since 1969. In September 1993, he went to a barber for the first time in 30 years—he had been wearing his hair in a ponytail and periodically hacking it off at the ends.

Yet though indifferent to fashion or fads, Dr. Conway is intensely aware of nature, and wants to know it deeply and intimately. He is infinitely curious and observant, seeing nature not only in a spiderweb or the details of a daffodil, but in mathematics. Part of his way of getting to know the world is through an eclectic mixture of problems, ranging from the logic of quantum mechanics to the best way to pack eight-dimensional spheres, to devising an aesthetically pleasing design for an artist who is going to plant a mountain with trees and wants to be sure that the design remains intact when trees are removed as the forest matures.

Dr. Conway was born and reared in Liverpool, England, the son of a laboratory assistant at the Liverpool Institute for Boys, a school that was attended by Paul McCartney and John Lennon of the Beatles. His talents in mathematics were always apparent. "My mother used to say she found me reciting the powers of 2 when I was 4 years old," he said. "I don't know if it's true and I don't know how far I got. It may be just one of those motherly stories." When he went to school, he said, "I was universally good at everything; I was the top or second or third in my class in every subject."

Yet mathematics always had a special appeal for Dr. Conway, and not just because he was good at calculations. "What turned me on was this mysterious relationship between things," he said. "There is this wonderful world of logic and connections that is very difficult to see. I can see trees and cats and people, but there is this other world and it's very, very powerful."

He explained that mathematical logic, the sort of reasoning that ties abstract ideas together, fascinated him and made him wonder how real it all was. Were abstract mathematical constructions illusions or reality? Were they a creation of the human mind or were they describing the essence of reality? Is mathematics discovered or invented?

Dr. Conway, along with many other mathematicians, believes that mathematics is discovered because, over and over again, abstract mathematical ideas, found purely by thought and reasoning, have turned out to provide exact descriptions of the physical world.

For example, he has been doing research with Dr. Neil Sloane of Bell Labs on sphere packing, which involves figuring the best way to pack spheres into a particular space. The work would seem to be an abstract and useless exercise. But it turns out that the eight-dimensional sphere packing problem is exactly what is needed to find the best way to transmit data, like the dimensions of a sound wave made by a human voice, along telephone lines. The idea is to take every eight data points, which name a point in eight-dimensional space, round them off, and transmit the rounded point. The rounded point is the center of the nearest eight-dimensional sphere.

"I find it really lovely that this purely geometrical thing that I'm interested in is actually useful to quite practical people," Dr. Conway said.

Dr. Conway and Dr. Sloane recently wrote a book about their sphere packing work, which was reviewed in the journal *Advances in Mathematics* by Dr. Gian-Carlo Rota of the Massachusetts Institute of Technology. It went, in part, like this: "This is the best survey of the best work in the best fields of combinatorics written by the best people. It will make the best reading by the best students interested in the best mathematics that is now going on." Dr. Conway calls it the "best" review.

Dr. Conway got his start as a renowned mathematician in 1969, when he was a faculty member at Cambridge University, hoping desperately to make a mark on the world. He was married and had four little girls. "I knew that I was a good mathematician, but the world didn't," he said. "I was getting very depressed because I hadn't lived up to the promise."

Another mathematician announced that he had discovered a new mathematical object, a highly symmetrical creation, in 24-dimensional space. By studying this, mathematicians hoped to find a new group, the concept they use to understand symmetry. But the discovery was preliminary and required working out the rules and symmetries to establish the group. Dr. Conway decided to make the attempt.

"I had this great plan," he recalled. "We were very poor at the time and I was doing lots of teaching to make ends meet. My plan was that I would work on it on Wednesday evenings from six until midnight and on Saturdays from 12 noon until midnight. I told my wife that this was very important."

He started work on a Saturday. "It was very funny," he said. "Basically, I said goodbye to my wife, kissed the kids and stuck myself in the front room." Twelve and a half hours later he emerged, to his own astonishment, with the task completed.

That was the turning point in Dr. Conway's career. "It catapulted me into the jet set," he said. Soon he was traveling around the world, much in demand to describe his work. "I remember flying to New York to give a 20-minute lecture and then flying back again," he said.

The discovery "had a great psychological effect," Dr. Conway said. "It cured this depression and it also totally removed my ambition. I've been totally successful, but at some level I couldn't care less." He used to dream of being like most of his colleagues who "think of very deep things," he said. But with his momentous proof of the existence of what has come to be called the Conway Group, "that sort of removed that feeling." He added, "I decided that I might as well enjoy myself instead."

So what does he do to enjoy himself? For one thing, he indulges in his fascination with symmetries. He studies the petals of a daffodil, figuring out why there have to be an alternation of three true petals and three leaflike petals called setals. "I feel like I understand it," he said. "It gives me a nice feeling of being in tune with the world. It's a feeling of harmony."

Or he decides to memorize the first 1,000 digits of pi. This was a project that his second wife, Dr. Larissa Conway, a Russian-born mathematician with whom Dr. Conway has two little boys, instigated when one day she needed the value of pi and knew it only as 3.14.

"I taught her the first 100 places, but she wanted more," Dr. Conway said. "I discovered I didn't know any more," he said, so he and his wife made it a project to memorize 100 digits a day. The two would practice together during their regular routines. "It was really rather sweet," Dr. Conway said. "We used to go on a walk on Sundays to the village and one person would recite 20 digits as we walked along, like a little poem, then hand it over to the next person and they would take over for the next 20 places."

He went on: "As I'm getting older, I'm finding that rather than having a deep understanding of lots and lots of problems, I like just knowing lots and lots of things. I don't mean facts—I mean things. Knowing what a daffodil looks like or the 633rd digit of pi. Whatever it is. Some are show-offy. Most have no earthly use. I just like the feeling, and it is of some use to me professionally."

Asked how he actually does research, Dr. Conway describes a life of the mind that to most experimental scientists would seem unbearably frustrating. "What happens most of the time is nothing," he said. "You just can't have ideas often." When he does not know what to try to prove, he says, he will just start enumerating something. "I might try and list all the polyhedra with nice properties," he said. "With any luck, I'll see something."

It is a method that gets him through the dry days, he said, and one that eludes most young researchers. "One of the terrible things that happen to bright young graduate students is they'll come in at 9 A.M. and write 'theorem' at the top of their paper. Then they might write 'proof.' Then they sit there staring at this thing, they get depressed, they go home at 5. They come in the next day and start again." The problem, he explained, is that "there's no way of proving a theorem to order. If someone comes up with a gun and says, 'Prove this theorem,' well, then, you're just dead."

A typical day for Dr. Conway starts at about 8 A.M. when he goes to a local coffee shop and reads a newspaper over a breakfast of oatmeal. Then he turns on his computer, works out 10 dates and reads his e-mail. "There might be a math problem in the e-mail that might make me think," he explained. He will often have very long telephone conversations with Dr. Sloane about the work they are doing together or he might go to Bell Labs to work. He is teaching two classes this semester.

"The point is, I don't work," Dr. Conway said. "I mean, honestly, if you or your readers saw what I actually did, they'd be disgusted. They'd say, 'Good money is being paid out to support these people!' "

Dr. Conway said he used to feel guilty about the way he worked. "Then it suddenly dawned on me that that's the way I produce good stuff—by lying around, doing nothing, thinking about what I like," he said. And, he added, "I've been a lot more successful since I stopped feeling guilty."

—GINA KOLATA
October 1993

ART, SCIENCE AND THE TRUE NATURE OF MATH

When mathematicians speak frankly about their discipline, they tend to circle around a fundamental question: are the ideas of mathematicians discovered or invented?

Mathematicians speak of proofs as "elegant" or "beautiful," praising work for its aesthetic nature. Dr. Mina Rees, a logician, wrote, "Mathematics is both inductive and deductive, needing, like poetry, persons who are creative and have a sense of the beautiful for its surest progress."

But, even so, most mathematicians say that when they do their work they are making discoveries—the great ideas of mathematics exist independently of them. They stumble upon these ideas, awkwardly, perhaps, but eventually finding truths of nature.

Dr. John Conway, for one, cannot imagine he is doing anything but discovering results that exist without him and that he did not create. Why? "Because they couldn't be otherwise than what they are," he said. "Two and two might be five and pigs might fly. But in the world I come from, two and two are four and pigs don't fly."

Yes, sometimes the world that Dr. Conway and other mathematicians say they are discovering can have infinite dimensions or can involve space that twists and turns and has holes in it. Sometimes the words sound like English—"fiber bundles" or "modular forms"—but there seems to be no English equivalent that would allow mathematicians to explain what it is they are talking about. Nonetheless, Dr. Conway said, "mathematics is about the real world."

Dr. Herbert Robbins, a statistician, has said that he spends most of his time filling his wastebasket with crumpled pieces of paper, but that when he finally sees how to solve a problem, he feels he is dimly grasping something that exists without him and that is almost beyond his capabilities. "When something significant is happening," he said in a 1985 interview, "I have a feeling of being used—my fingers are writing but it's hard for me to get the message."

Mathematicians love a notion described by the late Hungarian mathematician, Dr. Paul Erdos, that there was a book, kept by God, of all the most beautiful mathematical proofs. Every problem mathematicians would ever encounter was solved in The Book, and in the most elegant possible way. The job of humans was to try to discover the writing in The Book.

But not everyone thinks the discovery-or-invention question has a single answer. Some say that they can discover one day and invent the next, and that the two process are distinct and distinguishable. Dr. Kenneth Ribet, a mathematician at the University of California at Berkeley, says that for him, it all comes down to the feeling he gets when he works.

With discovery, Dr. Ribet said, "you feel like you are on a path." He said he may try one approach after another trying to solve a problem, but when he discovers the right one, it is unmistakable, like being lost in the woods and finally finding the path that leads you back home. "It's like, 'Aha, I found the right thing,' " Dr. Ribet said.

Creation involves very different feelings, Dr. Ribet said. He uses it when there are no tools to solve a problem, and so he invents something, a notion that may or may not work but that clearly, he said, did not exist until he dreamed it up, giving it shape and form. The feeling, he said, "is like being an artist."

And along with the artistic impulse comes the artistic despair.

"If you're in a good mood, you call yourself an artist," Dr. Ribet said. "If you're in a bad mood, you feel unworthy and overwhelmed." His inventions, he said, "are very tentative." And they need to be subjected to tests. "Only after time can you see that it is the right thing."

—G.K.

Martha K. McClintock

How Biology Affects Behavior, and Vice Versa

Steve Kagan *for* The New York Times

One of the enduring clichés about scientists is that while they may be a bit large for the average seesaw, and indeed may even have graying beards, they remain forever young at heart: full of childlike wonder and curiosity about the natural world, their eyes as round as pies, their energy infinite, their speech a variant on the worshipful "Wow!"

Dr. Martha K. McClintock, a professor of biopsychology at the University of Chicago, is no child, and her sense of curiosity is too richly alloyed by years of research, skepticism, point and counterpoint to fit into any child's brain. Yet one thing about Dr. McClintock, 47, is distinctly kidlike: she loves to make a mess.

Most researchers who study animal behavior in the laboratory keep their subjects in standard-issue cages and try to minimize the confounding variables, to zero in on one behavior isolated from all distractions. Dr. McClintock does just the opposite. When she puts animals into an enclosure, she deliberately turns the place into a dump.

Showing off the room that has been established for her lab's collection of Norway rats, she declares proudly that it is just the sort of environment where the rodents feel most at home; Norway rats, it must be pointed out, are the familiar dun-coated, naked-tailed creatures that thrive in landfills and alleys and sewers. There is litter all over the floor, and plastic tubes and nooks and tight places for the rats to slither through, and burrows for breeding. "Some of my students think there should be smashed hubcaps and grain silos in here as well," she said. She apologizes because no rats are running loose in the room at the moment, and the apology is graciously accepted.

Dr. McClintock is currently overseeing the construction of her long-term dream, a $12 million institute devoted to the study of the inextricable link between environment and biology—two sides of the same Möbius strip, as it were. She shows off blueprints for the planned building, designed by one of Chicago's most prestigious architectural firms, Holabird & Root. The design of the exterior looks sedate, prim, boxy, and she is now trying to persuade the architects to mess it up a bit. She wants something more "aesthetic," more Gaudi-esque. "I keep telling them science is not square, science is organic," she said.

Above all, Dr. McClintock is committed to the concept that life is inherently messy, complicated, cross-wired, an endless negotiation between interior and exterior, between the environment in which a human or an animal lives and its physiological and neurological state. She argues that the vast majority of research has focused on how biology affects behavior and has neglected the equally important question of how behavior affects one's biology.

"If you look at the journals in my field, 90 percent of the articles look at the effects of physiological, neural and hormonal systems on behavior, and 10 percent look at the effects of behavior on hormones and the nervous system," she said. "I don't think a balance of 90-10 is an accurate reflection of how nature works."

As an example of how behavior can sway physiology, she points to studies in stress research, in which scientists have sought to understand what distinguishes a person subject to bursts of stress hormone activation from one whose cortisol levels are less easily aroused. They have found that an essential factor is a matter of opinion—whether or not one believes one is in control of a situation. The person may or may not actually be in control, but simply feeling as much profoundly affects one's endocrine system.

"How can something as amorphous and difficult to localize in the brain as an opinion or belief trigger something as concrete as a change in cortisol release?" she said. These are the sorts of questions that she feels have been relatively ignored in recent years, and that she intends to pursue in her new institute.

That a university is undertaking to build an expensive new research facility in an era when biologists everywhere are seeing their grants hacked to the marrow is a credit to Dr. McClintock's large and long-standing reputation in the field of biopsychology—where the body meets the brain and the pair meet the world—a reputation that began in her graduate school days. Her instructor at Harvard University, the celebrated naturalist Edward O. Wilson, persuaded her to publish in the journal *Nature* data that she had collected as an undergraduate at Wellesley College, showing that women living together in dormitories eventually ended up with synchronous menstrual cycles. The paper was wildly successful in the scientific and popular arenas alike, prompting Dr. McClintock to a broader consideration of how signals from one's surroundings may get translated into bodily changes.

Among her many research projects now are studies of how female rats coordinate their reproductive cycles to allow them to give birth to pups en masse, allowing for communal nursing pools and healthier offspring. As it turns out, the female rats generate airborne chemical signals called pheromones that either enhance or suppress the fertility of their neighbors. This rodent synchrony of fertility offers an animal model for understanding why and how such group cycling occurs. Her lab has also discovered that when female rats are prevented from interacting communally with other females, they go into the rat's version of menopause comparatively early and are at heightened risk of breast cancer and other cancers. Dr. McClintock suggests that this recent work may give insight into why social isolation among humans increases one's risk for an early death—again, an example of how a particular behavior can rewrite the biological script.

"Her research has always been excellent and she's come up with some very important ideas," said Dr. John Vandenbergh, a zoologist at North Carolina State University in Raleigh. "It's also essential that someone like her moves back and forth from human studies, which are largely correlation and observational, to animal studies, where can you pin down the mechanism to understand it all."

Also behind her success in getting the institute under way was her obstinacy in the face of improbability. "I've been at this for a decade," she said. "I would not go away." Obstinate does not mean obnoxious, however. Dr. McClintock has the sort of good-humored, sororal and charismatic personality that draws people to her; when her students were asked how they liked working with her, suffice it to say that they gushed. Dr. McClintock is tall and slim and dresses so stylishly that even her socks are interesting.

The university was also convinced of the need for the institute, a place where the focus will be on a top-down approach to understanding biology—going from the whole organism toward the innermost layer of the gene—rather than the standard bottom-up, beginning with the gene and looking outward. Genetics is currently in the ascendancy, what with the huge Federal Human Genome Project that seeks to understand the human genetic blueprint. Even at the National Institute of Mental Health the emphasis is on genes above conventional social science and behavioral studies.

While conceding the enormous power of modern genetics, Dr. McClintock insists that the current fixation on genes and molecules above all else is a shortsighted and ultimately an impoverished approach to understanding many of the most alluring problems in biology, particularly questions of human behavior. "There are people who are pragmatic reductionists," she said, referring to those who study molecules because the tools exist, yet who do not necessarily believe that they hold the whole truth of how nature operates. "Then there are people who are, in principle, reductionists," she said. "They think that complex problems like behavior, a whole person's interaction with the environment, feelings, wishes and memory, can ultimately be analyzed at the cellular and molecular level.

"I argue that, in principle, it's not true," she continued. "The theme underlying everything we'll look at in the new building is to provide balance to this reductionist bias, this focus on getting ever more molecular."

Dr. McClintock calls the molecularization of science "falling through the membrane and forgetting to climb back out."

Dr. McClintock sees many instances where a genetics slant alone reveals nothing. For example, she is particularly excited about one project in her lab that focuses on how female rats control the number of male and female offspring they bear. As it happens, mother rats give birth to very different ratios of sons and daughters depending on how they are faring. When the rats are doing well and mating with vigorous, dominant males, they give birth to an excess of sons. When times are hard, or when they have recently lost a litter, they give birth to more daughters.

Daughters, it seems, are the "cheaper" sex to raise: mothers spend less time licking and nursing them, and they are smaller than sons at weaning. They are also the safer sex, almost assured of having some offspring and thus keeping their mother's genetic legacy alive. Sons, by contrast, are considered the jackpot sex, who in theory could do brilliantly by their mother come their sexual maturity, spawning far more offspring than their sisters ever could manage. Hence, there is evolutionary justification for mothers to bear more expensive sons when they have the energy and resources to do so, and opting for fail-safe females when prospects are dim. Somehow, the mothers extract relevant information from their habitat and translate it into a changed birth ratio.

"It's mind-boggling to figure how these social cues regulate something"—the sex of one's offspring—"we think of as completely genetic," she said. The early results suggest that the mother makes her unconscious "decision" about how many of each gender to bear before the embryos become implanted in the womb, and the researchers are trying to track down the mechanism behind that selective implantation.

Dr. McClintock anticipates that perhaps eight professors from the University of Chicago and elsewhere will end up moving into the new building. They will probably include Dr. Jeanne Altmann, a renowned primatologist who studies yellow baboons and the hormonal correlations of their struggles over rank and status; Dr. Stevan J. Arnold and Dr. Lynne Houck, who have recently discovered salamander mating pheromones; an immunologist who will look at the influence of behavior on the immune system; a neurobiologist to look at the neural mechanisms that control well-defined behaviors, the better to explore how plastic and responsive to the environment those mechanisms may be; and a molecular biologist specializing in the genetic aspects of behavior.

The interdisciplinary texture of the enterprise suits Dr. McClintock's temperament. Her degrees are all in psychology, but she has worked in evolutionary biology and neurobiology and did a residency in psychiatry at the University of Pennsylvania School of Medicine. She stays linked to medicine in part because she is married to a physician, Dr. Joel Charrow. They have two children, Benjamin, 9, and Julia, 2.

As much as Dr. McClintock insists on the suppleness and tractability of biology and behavior, she does not believe that people emerge in the world tabulae rasae, able to write whatever plot they choose. She says she breaks with some feminists in believing the scientific evidence suggests that a few behavioral differences between men and women are innate. The differences are small, and they apply only on average, she said, "but they are there and they're real and they're not just the product of social construction."

As an example, she cited studies indicating that men, on average, are better than women at focusing on the task at hand and ignoring the extraneous noise around them. That conviction led her to agree recently with an Air Force officer that while there are many excellent female fighter pilots, the natural ratio of males to females in the profession might not be 50-50.

On the other hand, she said, women have been shown to surpass men in tracking the multiple signals that are going on in a group—whether it is at a social gathering, in the boardroom or at the negotiating table. Women are much better, she said, at "integrating the words, the intonation, the facial expressions, and figuring out what people's ulterior motives are." Which suggests that, if nature were allowed to take its course, one might expect women to outnumber men in politics, diplomacy, trade negotiations and a host of overwhelmingly masculine fields.

—NATALIE ANGIER
May 1995

WHERE GENES ARE JUST PART OF THE STORY

When last we met Martha McClintock, she was busily making a mess of things, turning lab rooms into ad hoc dumpsters so that the Norway rats her team was studying would feel at home and behave accordingly.

Since then, she has cleaned up professionally as few scientists can hope to do in their lifetimes. She is the director of a new $16 million institute at the University of Chicago devoted to a subject dear to her heart, mind and the many conduits that join them: the link between biology and environment, physiology and context. In 1995, the institute was as yet prenatal. It didn't have a name, not all of its funds were accounted for, and the preliminary architectural plans were, in Dr. McClintock's words, "plain and awful."

Now, the institute is built, up and running, with all the manic energy of a rodent and the calm beauty of a Rodin. Designed (and redesigned) by the celebrated firm of Holabird & Root, the completed structure is "quite spectacular," said Dr. McClintock. "It's got every feature you'd want to encourage interactive science, and it's really gorgeous besides." It has an official title: The Institute for Mind and Biology. And it has a growing staff of scientists devoted to the concept that nature abhors an arrow and adores a loop.

Whereas the common perception has it that genes, hormones and brain structures unilaterally dictate how an individuals looks, feels and behaves, the researchers at Dr. McClintock's institute argue that one's environment and social setting can feed back to the body, altering an individual's chemical state, the contours of the nervous system, even the activity of DNA itself.

With that in mind—and body—Dr. John Cacioppo studies how loneliness and a sense of existential anomie can disturb the body's autonomic nervous system, which controls many automatic functions like heartbeat, digestion and blood pressure; and, on the flip side, how feeling socially "embedded," or beloved and wanted, enhances the performance of a person's autonomic nervous system and all parts attached thereto.

Dr. Dario Maestripieri, a primate researcher, explores how the emotional bond between a mother and her baby affects physiology, health and well-being—not of the infant, which most scientists concern themselves with, but of the mother.

For her part, Dr. McClintock and her co-workers continue to study human pheromones, chemicals emitted by one person that influence the behavior or emotional state of another. In a widely acclaimed report in 1997, they presented evidence that women release odorless chemical signals that can either shorten or lengthen the menstrual cycles of other women in their vicinity.

More recently, the scientists have examined two steroid chemicals, isolated from human skin, which several enterprising perfume manufacturers have blended into their fragrances and touted as sex-specific pheromones. One compound, androstadienone, is added to a woman's perfume, purportedly to make her feel sexy and self-confident, and her partner feel enticed. Another extract, estratetraenol, is added to men's cologne, with the goal of a similar effect on women.

The McClintock researchers found no evidence for the perfumers' exalted claims, but they have shown the chemicals to influence people's emotions. It seems that both the "male" and "female" extracts exert a similar impact: they improve the moods of most women exposed to them, and they worsen the tempers of most men.

Poor fellows: there's just no satisfying them, is there?

—N.A.

Preaching the Gospel of Healthy Hearts

Rick Friedman for The New York Times

Dr. William Castelli, the director of the Framingham Heart Study, was not always renowned for his bons mots and memorable quotes, which have helped make the concept of "good" and "bad" cholesterol familiar to millions of Americans and prompted them to recast their diets and exercise habits to save their hearts.

He remembers all too well his first professional lecture, when he tried to demonstrate the role of a new cholesterol-lowering drug to a group of general practitioners. Scrawling away on a blackboard with his back to the audience, he painstakingly detailed all 26 steps in cholesterol metabolism. But when he turned on the last step to deliver the punch line about the drug's action, he found the entire audience, including his wife, sound asleep.

In more recent years, he has been called by various professional colleagues "a very enlightening, entertaining and effective lecturer," "a great communicator who talks to the public in meaningful sound bites" and "a man with a real passion and very unique ability to communicate what he finds through scientific observation." One of his rare detractors paid him a backhanded compliment by dubbing him "the Oral Roberts of prevention." He is constantly invited to lecture to professional and lay audiences around the world, spreading the gospel on how to save hearts and health.

It is a subject he may know better than anybody. As a 28-year veteran at Framingham, and its director since 1979, he can rattle off findings gathered through decades of measuring the health characteristics and life habits of more than 10,000 Massachusetts townspeople and matching these with their eventual medical fates.

Dr. Castelli also relishes a good fight and is often invited to debate the naysayers and skeptics about coronary risk factors. Using facts and figures, most of which have been established by the 45-year-old Framingham study, he readily reduces their arguments to ashes. Despite a certifiable Type B personality that seems never to become agitated, "he goes for the jugular," remarked Dr. William Kannel, the former director of the study, who hired Dr. Castelli in 1965 to run the Framingham laboratories. "Many are no longer willing to debate with him."

But, Dr. Castelli says: "Teaching and lecturing did not come naturally to me. It took a lot of hard work."

Hard work has clearly paid off, as is evidenced by his popular "hot dog lecture." He recently gave an abbreviated version. "How do you make a hot dog?" he asked. "First you slaughter the animal and cut out all the good parts, the steaks and chops. But you've got a lot of animal left and what are you going to do with it? The hot dog industry took off when a clever guy invented a machine that works like a kitchen disposal—you dump everything in, eyeballs and all, and grind it up. Voila, the hot dog."

His unorthodox lecturing technique also includes taking students on a tour of a typical American supermarket, with Dr. Castelli flipping over one product after another to show from the fine print on the labels how American arteries become clogged with the fatty deposits that have made heart disease overwhelmingly the leading cause of death in this country.

On a recent supermarket venture, Dr. Castelli took a slightly different approach and decided to show how health-conscious consumers can have their cake and eat it, too—eating the kinds of food they want without burdening their blood vessels with cholesterol-raising saturated fats. By choosing wisely, for example, buying "special lean" ground beef with less than 10 percent fat instead of the regular beef with about 25 percent fat, he said, "theoretically, you could have a 4-ounce hamburger for breakfast, lunch and supper and only use up 12 grams of the day's allotment of 20 grams of saturated fat." The low-fat burgers have an added advantage. "They don't sizzle or flare up and make a mess of the stove," said the sometime cook, who is as adept in the kitchen as he is before an audience.

"I used to tell people, 'When you see the Golden Arches, you're probably on the road to the Pearly Gates.' But I stopped saying that after McDonald's took the beef tallow out of its fries and gave us the McLean burger, with only 4 grams of saturated fat."

Moving down the meat counter, Dr. Castelli found a package of skinless sliced chicken breast advertised as 97 percent fat free and then moved on to a package of sliced ham that was 98 percent fat free, which means that it has about 1 percent saturated fat, less than half that found in the lowest-fat ground beef.

"You like ice cream?" he asked. "Well, one cup of Ben and Jerry's ice cream has about 35 grams of total fat and 20 grams of saturated fat." Instead, he picked out a carton of low-fat frozen yogurt from the same manufacturer, saying, "you'll get only 4 grams of fat and 2 grams of saturated fat."

With his supermarket cart now laden with more than a dozen low-fat and fat-free products to sample, he summed up: "People could control their destiny. They don't have to go on to get some degenerative disease."

It has been said that a heart attack is a good, clean way to die. Dr. Castelli disputes that notion. "Most people don't die after their first heart attack or stroke," he said. "They go on to angioplasties, bypass surgery and fistfuls of medication, and are sick for the rest of their lives."

Dr. Castelli noted that half of all heart attack victims lose the normal pumping ability of their hearts and suffer from fatigue, swelling and shortness of breath. "A heart attack takes the octane out of your tank," he said. "You can't shovel snow, can't do the tango, can't pick up a suitcase or your grandchildren."

To those who think their decades of bad habits make it too late to change, Dr. Castelli counters: "We now believe it's the newest lipid deposits—where that greasy cheeseburger you just ate landed—that rupture and precipitate the majority of heart attacks. These deposits don't impinge on blood flow while on artery walls, but they are unstable and easily rupture to block a coronary vessel. The good news is that these young unstable deposits can be shrunk. Even if you have a heart attack, after two years of lowering your cholesterol, you can dramatically reduce your risk of another attack. There have been 18 or 19 studies on reversibility, and they all show that if you really get the numbers down, the lesions in your arteries start to shrink."

He laments that Americans are a long way from "aggressive lipid management," which would enable 75 percent of them to lower their cholesterol through diet and exercise alone to a level where a heart attack is highly unlikely. The remaining 25 percent would need the help of cholesterol-lowering drugs, he said, adding that only 5 percent are currently on such medication.

As the first male in his family to make it to 45 without coronary symptoms, Dr. Castelli considers himself living proof of the potential of changing to a healthful lifestyle. His father, who was a corporate physician in New York, lost a leg in his mid-50's because of clogged arteries and his brother developed angina in his early 40's. "But I've now made it to 62 without a heart attack," he said, despite the fact that until his mid 30's his diet was rich in all sorts of fats and cholesterol.

He achieved this not by becoming a fanatic but by watching "very carefully" his intake of saturated fats to lower the "bad" cholesterol carried in the blood by LDL's (low-density lipoproteins) and by exercising regularly to raise the "good" cholesterol carried by HDL's (high-density lipoproteins).

"In my late 20's and 30's my cholesterol level averaged 260 or 270," he recalled. "In those days it was considered bad if your cholesterol was over 300. Then, in 1965, I came to Framingham to reorganize the study's laboratories and start measuring the participants' lipoprotein levels. I discovered that the average man in Framingham who had a heart attack had a cholesterol level around 240. I said to myself, 'How come my cholesterol level is considered normal when half the heart attacks are occurring at levels below 240?' So I went on a prudent diet and got my cholesterol down to 200."

Fish and shellfish are regular features of his heart-healthy diet because they are naturally very low in saturated fat. "If you can't be a vegetarian yourself, the next best thing is to eat a vegetarian from the sea," he said. These are the mollusks—mussels, clams and oysters—"the animals that are rock-bottom lowest in saturated fat." He added: "Even the crustaceans—shrimp and lobster, for example—are better to eat than the white breast of chicken without the skin because they are so low in saturated fat. The cholesterol in crustaceans has been recently reanalyzed and found to be much lower than we used to think. You can eat two dozen shrimp and still take in only 200 of the 300 milligrams of cholesterol allowed in a day."

Then, in the late 1970's, the Framingham data revealed that even more important than one's total cholesterol is the relative amount of HDL cholesterol in the blood, which acts like arterial drain cleaner, clearing fatty deposits from the circulatory system. The higher the HDL's, the better. So Dr. Castelli started jogging and raised his HDL's from 49 to as high as 63.

Dr. Scott Grundy, an expert on dietary fats at the University of Texas Southwestern Medical School in Dallas, credits Dr. Castelli with bringing the importance of HDL's to the attention of professionals and the public. Although at first the National Cholesterol Education Program resisted emphasizing HDL's, measurement of these protective lipids has since become part of the initial testing recommended by the Federal prevention program.

Dr. Castelli insists that there is no substitute for "taking on an issue and studying it to its greatest depth." With all his preaching about prevention, he said he never goes beyond the facts that have been established by well-designed research.

As for the future, he says: "I've always wanted to teach epistemology, how you know what is true—how to consider all the evidence, where you get it, how reliable it is, and all the exceptions that allow you to come to the conclusion, such as that high cholesterol is bad for you. We don't teach that in college or in medical school." He believes that is why even well-educated people have so much trouble sorting out the facts about staying healthy.

Dr. Claude Lenfant, who heads the National Heart, Lung and Blood Institute, which supports the Framingham study, says, "Everyone thinks Framingham is Dr. Castelli," adding, "He would be very difficult to replace." But its director believes that Framingham now has a life of its own.

"It is a place that discovers, proves, establishes in an epistemological sense what are the risk factors for heart disease," Dr. Castelli said. "The findings of Framingham have already helped millions of people around the world, and even if the older generation is not helped directly, their children, grandchildren and great-grandchildren will be helped."

—JANE E. BRODY
February 1994

A TOWN OF WILLING HEARTS

For more than half a century, thousands of residents of Framingham, Massachusetts, have been living in a medical fishbowl. They are anonymous donors of an incomparable wealth of data that has helped researchers determine why some people stay healthy while others develop high blood pressure, diabetes, heart attacks and strokes. For going on three generations now, the people of Framingham have contributed more than the residents of any other community to advancing the health of Americans.

When the United States Public Health Service, alarmed at the soaring rate of heart disease among Americans, started the study in 1948, the harmful effects of high blood levels of cholesterol had yet to be documented. It took Framingham to show that what was once thought to be a normal, healthy level— 240 milligrams in 100 milliliters of blood serum—was associated with a high risk of heart attack. Half the heart attacks in Framingham occurred among people with levels of 240 or less.

Nor had clear-cut links between heart disease and smoking, high blood pressure, obesity, diabetes, menopause, a sedentary lifestyle and excessive levels of stress been established when the study began. In fact, it was Framingham researchers who developed the entire concept of coronary risk factors, which in the decades since have enabled physicians and public health officials to guide Americans towards healthier living habits, which in turn has resulted in an unprecedented decline in coronary mortality.

Prior to Framingham, doctors thought exercise was dangerous for people at risk of developing heart disease. However, studies at Framingham documented the preventive value of exercising, quitting smoking and taking medication to lower high blood pressure. The observations made in Framingham led to clinical trials that proved the life-saving value of reducing coronary risk factors.

The Framingham study has also yielded a host of surprising, and at times dismaying, findings. Among them: chronic high blood pressure can lead to mental declines beyond those associated with natural aging; significant weight gain after age 25 significantly increases the risk of heart disease in both men and women; rapid onset of baldness is associated with an elevated risk of developing and dying from heart disease; high blood levels of the amino acid homocysteine are linked to clogged arteries feeding the brain, which raises the risk of a stroke; and eating lots of vegetables and fruits, some of which help to lower homocysteine levels, curbs the risk of stroke.

Research on other diseases has also benefited from observations in Framingham. For example, no increased risk of breast cancer was found among women who consumed moderate amounts of alcohol—up to one and a half drinks a day. This was especially good news because that amount of alcohol lowers the risk of heart attacks. And, among elderly residents of Framingham, those who took supplements containing vitamin D were less likely to develop arthritis in their knees. As more researchers mine the riches of Framingham, findings should continue to emerge that can lead Americans down a healthier path.

—J.E.B.

JoAnn M. Burkholder

In a Sealed Lab, a Warrior Against Pollution

Karen Tam for The New York Times

Dr. JoAnn M. Burkholder looked into the television monitor attached to a powerful microscope. Before her, a swarm of tiny killers swam into view. "Oooo, look at all those," she said as the microbes darted by the dozen across a glass slide covered with human blood and attacked the red cells, sucking them dry.

One of them, twitching with what seemed like elation, stuck its proboscis into a blood cell and spun around in a gruesome dance. A rival quickly joined in.

"We haven't seen them fighting over one cell before," mused Dr. Burkholder, the head of a small team at North Carolina State University in Raleigh studying the microorganism. "That's really interesting."

No one knows better than Dr. Burkholder how bizarre—and how dangerous—this microbe can be. She discovered and named *Pfiesteria piscida*. She has seen it cripple rivers in huge fish kills, raising serious questions about its threat to fishermen and people who eat diseased fish. And she knows first hand its devastating effects on humans, having undergone a nightmare of nausea, disorientation and memory loss.

As a result, hers has been one of the loudest voices in both official and unofficial forums calling for a war on the microscopic killer and its deadly cousins in the waters of North Carolina, where *Pfiesteria* (pronounced fee-STEER-e-ah) is nourished by runoff from urban development and industries like hog farming. While many fishermen and environmentalists applaud her work, her uncompromising stand has won her the enmity of big financial interests often responsible for big pollution.

Her microscopic subject is one of the most deadly representatives of a diverse class of aquatic organisms that cause red and brown tides, watery scourges apparently on the rise globally. It can assume at least 24 guises in a lifetime of killing fish and shellfish in coastal estuaries. It also attacks humans directly in the lab, and apparently in the field, and clearly releases extremely potent toxins that stun fish and afflict humans who breathe the fumes.

Scientists call it "the cell from hell," and Dr. Burkholder, an aquatic ecologist, has linked its rise to riverine pollution, including human sewage, hog feces and agricultural runoff and other nutritive substances. But her calls for aggressive studies and strong anti-pollution steps have prompted little action. Hog farmers, state officials and other advocates of development have challenged her character and competence. Anonymous foes have made death threats.

Her efforts have brought her attention not only in North Carolina but now nationally, too. A book, *And the Waters Turned to Blood* (Simon & Schuster), by Rodney Barker, chronicles Dr. Burkholder's long battle to win recognition for the dangers posed by the microscopic killer.

An intense, blue-eyed woman of 43, Dr. Burkholder works in a number of laboratories at North Carolina State. One of them, housing the microbes in their most toxic form, is on the fringes of campus where the microbes are cut off from the outside world by a complex system of containment vessels and air filters. The lab for microscopic studies is more centrally located in the windowless basement of Gardner Hall.

Her office on the top floor was cluttered with books and empty diet Coke bottles and acerbic cartoons. "The state's working on it," says a woman gazing deadpan at a hog coming out of her kitchen faucet, "but I have a few ideas what the problem is . . ."

A long bulletin board held a dozen or so ticket stubs from rock concerts, as well as letters of thanks from low-level state employees.

"We are all so indebted to you," read one note from a worker at the North Carolina Department of Environment, Health and Natural Resources. "You are an inspiration, but at the same time I deeply resent the hoops you had to jump through to be heard."

A food and water tray was on the floor for her 2-year-old dog, who visits occasionally. The phone rang constantly. Dr. Burkholder happily let the answering machine handle the calls while she described her work, her history and her sense of outrage. Rummaging through a thick stack of papers on her desk, she pulled out a document.

"Here," she said. "Read this."

It was a recent petition from more than 70 North Carolina physicians worried that a rash of patients with festering sores were poisoned by river water awash in such things as *Pfiesteria*. They won a year-long moratorium on new hog farms in three counties drained by the Neuse River on North Carolina's eastern coastal plain.

In the past few years, the Neuse and other nearby estuaries have suffered repeated kills, episodes in which millions of fish died, their bodies covered with bleeding sores torn open by *Pfiesteria*.

Sows in North Carolina last year gave birth to 16.2 million piglets, just a whisker behind the nation's traditional leader, Iowa, which produced 17.5 million new pigs, the Agriculture Department in Washington says.

"A lot of times I would have liked to walk away from all this," Dr. Burkholder said, but felt she could not. "People are getting hurt."

Though she is a heroine to some, her detractors call her a zealot quick to push beyond facts in pursuit of questionable policies.

The director of the state's water quality program in 1994 scrawled an obscenity across cleanup recommendations that a state advisory panel put forward. The panel's membership included Dr. Burkholder, its only scientist.

Now, in the glare of publicity surrounding the fish kills, some North Carolina state officials are denying any misfeasance and voicing strong support for Dr. Burkholder.

"I consider her a fighter for what is right and an outstanding scientist," Dr. Ron Levine, the state's health director, said in an interview. "This alleged conspiracy to try to suppress the work is wrong. We want to know the answers. This is an extremely important issue."

Mr. Barker's book portrays Dr. Levine as working to belittle her research and limit her financing, both of which he denied.

Debbie Crane, head of public affairs for the state's Department of Environment, Health and Natural Resources, said: "Bureaucracies by their nature are slow, which is unfortunate. We have not dealt with this situation perfectly." Ms. Crane voiced "great respect" for Dr. Burkholder.

That opinion is echoed nationally. Dr. Jane Lubchenco, an ecologist at Oregon State University who heads the American Association for the Advancement of Science, the world's largest federation of scientists, heaped praise on Dr. Burkholder.

"She's a very good scientist and tremendously courageous," Dr. Lubchenco said. "She thinks about things in ways that are different, and that's one of her strengths."

Politically, Dr. Lubchenco added: "She's gotten in trouble because she hasn't played the game by the old-boy rules. There's lots to be said for being politic and changing the system from within. She does that only so much, and then she rebels."

JoAnn Marie Burkholder was born in Rockford, Illinois, in 1953, the younger of two girls.

She fell in love with the outdoors and eventually won degrees to advance her ecologic interests—a bachelor's in 1975 from Iowa State in zoology, a master's in 1981 from the University of Rhode Island in aquatic botany and a doctorate in 1986 from Michigan State University in botanical limnology, the study of fresh waters.

She landed a job in 1986 at North Carolina State University, a sprawling land-grant school founded almost a century earlier. She had wide freedom to pursue whatever research interested her, and that turned out to be *Pfiesteria* and the trouble it was causing in the state's waters.

Dr. Burkholder stumbled on *Pfiesteria* when a colleague's laboratory research fish kept dying mysteriously. She found, studied and named the remarkable culprit, and by 1991 linked it to native North Carolina waters and extensive fish kills.

No one knows where it came from, but Dr. Burkholder in her research has discovered that *Pfiesteria* takes at least 24 forms as it moves from river bottoms to midwaters to feed on fish. It lives as far north as the Delaware Bay and as far south as the Gulf of Mexico, and may turn out to be ready to thrive around the world, given the right conditions.

She learned its devastating power at first hand in 1993 upon breathing toxic fumes from laboratory tanks of dying fish. The resulting nausea, burning eyes and cramps were bad enough, but then she began having trouble remembering phone numbers, writing or even holding a conversation. The fumes had crippled her memory and mental powers.

Howard B. Glasgow Jr., her chief research aide, was nearly overcome and ended up crawling out of the facility on hands and knees.

While her own recovery was relatively quick, his took months. And both researchers still complain of lingering effects, including an inability to exercise strenuously without having shortness of breath and the onset of respiratory illnesses.

Today, Dr. Burkholder's political activism has a steely edge, although she can still express sympathy for her early critics.

"It's no wonder that people were skeptical," she said in her office. "I had no credentials. No one had heard of me. Added to that was the problem that the organism is so bizarre. The backdrop to this story is the biology of a very fascinating organism that many people found hard to believe."

Today, after years of false starts, *Pfiesteria* research is on the rise. Money is coming from both state and Federal authorities, including the National Science Foundation and the Environmental Protection Agency. The work aims at finding out what causes the organism to transform itself into a killer (in the absence of suitable prey it can masquerade as a plant or lie dormant for years), and the exact makeup of its toxin.

"Only when that's characterized can we tell people if it's safe to eat fish; only then can we do a good epidemiologic study to see how people are being affected," Dr. Burkholder said.

For Dr. Burkholder, who first turned her attention on the tiny killer eight years ago, the investigation is only beginning.

"I'm worried about the insidious things, about the possibility of immune suppression from chronic exposure," she said. "Maybe children and elderly people are more susceptible. Right now we don't know.

"All I asked for through all this was for our state to seriously and responsibly consider this issue," she added. "There are many unresolved questions. But there's enough compelling evidence that the state should have, four years ago, begun a serious look at the issue. That still hasn't happened."

—WILLIAM J. BROAD
March 1997

HUMANS FIGHT BACK

The toxic scourge that Dr. JoAnn Burkholder identified and linked to extensive fish kills and human sickness has spread over the years, even as wide precautions inspired by her analysis are also spreading and having some success at beating back the cell from hell. Victories in the war, ecologists hope, will increase as more ambitious plans take effect to cut the pollution that feeds the microscopic killer.

Pfiesteria piscida hit the Chesapeake Bay region particularly hard during the summer of 1997, leaving perhaps millions of fish dead or thrashing weakly, many covered with gaping raw sores. The epidemic prompted Governor Parris N. Glendening of Maryland to order the closure of three of the bay's Eastern Shore tributaries to fishing and recreation.

Later the same summer, representatives of six states, backed by Federal environmental and public health agencies, declared an allied front in broadening the war against the toxic microbe. At the meeting, the governors of Maryland, Virginia, Delaware and West Virginia, as well as representatives of North Carolina and Pennsylvania and the Clinton Administration, agreed to share data, conduct joint research, and look for ways to better control polluted runoff of nutrients that may encourage the microbial blooms.

The Federal Government in late 1997 unveiled a new approach to battling water pollution today: planting trees to control runoff from farm fields. Agricultural runoff had long been seen as the soft underbelly of the Federal Clean Water Act. And in Maryland, some of the worst *Pfiesteria* outbreaks had been linked to the widespread use of chicken manure as fertilizer on nearby corn and soybean fields.

Scientists say that when land beside streams is used for growing trees, shrubs and grasses instead of row crops like corn and beans, the resulting buffer zones can reduce the amount of waste runoff by as much as 90 percent. Maryland and the Federal Government said they would spend up to $200 million to have farmers plant such buffer zones along practically all the waterways that flow through farms into the delicate Chesapeake Bay estuary.

The regional efforts inspired chicken and turkey producers across the country to try to develop a voluntary plan to handle the runoff from their operations, in hope of avoiding either costly new Federal rules or state-by-state rules that could create a competitive imbalance.

By the summer of 1999, the Chesapeake Bay was showing signs of better health. Improved water quality in some tributaries was linked to the regrowth of underwater grasses, the return of striped bass, a rebound in oyster populations, and the lack of major *Pfiesteria* outbreaks.

Unrelenting, the Environmental Protection Agency in the fall of 1999 moved to tighten things up still more, issuing a draft set of pollution guidelines for livestock factory farms. Still, experts agreed that much work remained to be done to curb the human sewage, hog feces, agricultural runoff and other nutritive substances that feed *Pfiesteria*.

—W.J.B.

Kary Mullis

After the "Eureka," a Nobelist Drops Out

Kary Mullis, Nobel laureate in chemistry, is jumping up and down at the kitchen table of his cabin in the woods of Anderson Valley in California, several miles beyond where the paved road ends. His large head and wiry body shake as if in rage. From his lips comes an angry buzzing sound.

He is imitating a swarm of yellow jackets, acting out an episode in which the wily insects ambushed him, inflicting five stings around the mouth, after he attacked their nest. He goes on to tell how he invented a brew, concocted in his kitchen blender, that eliminated the aggressors from his property for a season.

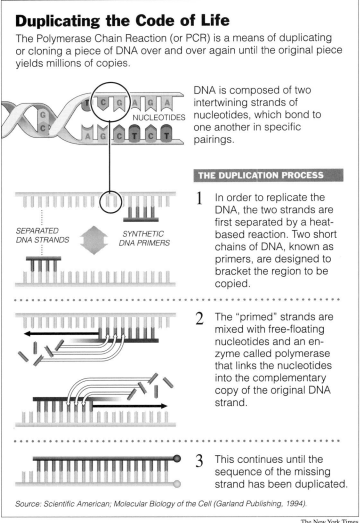

Duplicating the Code of Life

The Polymerase Chain Reaction (or PCR) is a means of duplicating or cloning a piece of DNA over and over again until the original piece yields millions of copies.

NUCLEOTIDES

DNA is composed of two intertwining strands of nucleotides, which bond to one another in specific pairings.

SEPARATED DNA STRANDS

SYNTHETIC DNA PRIMERS

THE DUPLICATION PROCESS

1 In order to replicate the DNA, the two strands are first separated by a heat-based reaction. Two short chains of DNA, known as primers, are designed to bracket the region to be copied.

2 The "primed" strands are mixed with free-floating nucleotides and an enzyme called polymerase that links the nucleotides into the complementary copy of the original DNA strand.

3 This continues until the sequence of the missing strand has been duplicated.

Source: Scientific American; Molecular Biology of the Cell (Garland Publishing, 1994).

The New York Times

He feels somewhat the same way toward his former colleagues at the Cetus Corporation, where he invented the technique that won him the Nobel Prize in 1993. "None of those vultures had anything to do with it," he says emphatically. He is aggrieved that Cetus paid him a mere $10,000 for the discovery but later sold it to Hoffmann-La Roche, owned by Roche Holding Ltd., for $300 million.

His invention, known as the polymerase chain reaction or PCR, is used for amplifying chosen sections of DNA and has quickly become an essential tool for biologists, DNA forensics labs, and almost anyone else who needs to study the genetic material. Amplifying DNA, a requirement for most tests, used to be done in bacteria, a process that took weeks. With PCR, performed with chemicals in a test tube, the job takes a few hours.

Science has been just one of the keen interests in Dr. Mullis's life, competing with psychedelic drugs and women, although he is now happily married to his fourth wife, Nancy Cosgrove. His newest interest is writing. A book of his essays, *Dancing Naked in the Mind Field,* was just published by Pantheon.

For those who would like to analyze creativity and sell it in bottles, Dr. Mullis would seem a promising subject. His invention is highly original and significant, virtually dividing biology into the two epochs of before PCR and after PCR. Yet the Mullis formula for creativity, on closer inspection, is a brew probably somewhat unsuitable for general consumption.

One ingredient is unbounded self-confidence. "Part of it has to do with his ego and belief that he's much smarter than the people around him," said Dr. Corey Levenson, a former Cetus colleague now at Ilex Oncology in San Antonio. "Most people who launch into an unfamiliar area would first speak to recognized authorities and get all the background. Kary saw that as a waste of time. He figured it would take less time to do the experiments himself."

Dr. Mullis's friends speak of his physical as well as intellectual risk taking. Dr. Frank McCormick, a cancer biologist at the University of California, San Francisco, recalls seeing Dr. Mullis in Aspen skiing down the center of an icy road through fast two-way traffic. "Mullis had a vision that he would die by crashing his head against a redwood tree. Hence he is fearless wherever there are no redwoods," Dr. McCormick said.

Along with lack of fear comes a lack of concern about people's opinions. In his book Dr. Mullis describes episodes that others might keep private, such as the time he addressed the Empress of Japan as "sweetie" when being awarded the Japan prize, and how he was nearly arrested when he went to Stockholm for his Nobel Prize, for playing a laser beam from his hotel room at passers-by.

His fondness for the heterodox is evident in the account of a lecture he gave in April 1994 at a medical society conference in Toledo, Spain. "Just before the lecture, he told me he would not speak about the PCR but would tell his ideas about AIDS not being caused by the HIV virus," the ambushed president of the society, Dr. John F. Martin, wrote afterward in a letter to *Nature.*

"His only slides (on what he called his 'art') were photographs he had taken of naked women with colored lights projected on their bodies," Dr. Martin continued. "He accused science of being universally corrupt with widespread falsification of data to obtain grants. Finally he impugned the honesty of several named scientists working in the HIV field."

Kary B. Mullis was born in 1944 in Lenoir, North Carolina, and grew up in South Carolina, where his father was a furniture salesman and his mother, who raised him after a separation, sold real estate. He trained as a chemist at the Georgia Institute of Technology and at the University of California at Berkeley. Then, shortly after getting his Ph.D., he dropped off the scientist's usual career path, first to write fiction and then, for two years, to manage a bakery.

It was a friend, Dr. Thomas J. White, who found him jobs back in science, first at the University of California, San Francisco, and then at Cetus in Emeryville, California, one of the first biotechnology companies. Dr. Mullis's job, essentially that of a technician, was to make short chains of DNA for other scientists. When machines became available to do the job, he had time on his hands for other pursuits.

Dr. Mullis has often described how the concept of PCR came to him during a night drive along Highway 128 to his cabin in Anderson Valley. He was playing in his mind with a new way of analyzing mutations in DNA and suddenly realized that he had instead thought up a method of amplifying any DNA region of choice. Before the trip was over, Dr. Mullis has written, he was already savoring prospects of the Nobel Prize.

The night journey was made in 1983; the Nobel Prize came 10 years later. But by then Dr. Mullis had dropped out of full-time science again. He left Cetus in 1986, earning his living by consulting and lecturing. He has published no more scientific papers. The divine spark that kindled the idea of PCR has not struck again.

"I like writing about biology, not doing it," Dr. Mullis says. "I don't want to go back to the lab myself and don't want to have people under my command. Fiction is my way around doing experiments."

He also enjoys giving lectures. "I love a microphone and a big crowd; I'm an entertainer, I guess."

Some of his agenda seems to have been selected with an eye to the shock value of adopting beliefs untypical of Nobel prize-winning scientists. He echoes the contrarian belief of the distinguished virologist Peter Duesberg that HIV is not the cause of AIDS. He disputes the arguments that chlorofluorocarbons are depleting the ozone layer and that industrial waste gases may cause the climate to get hotter.

"Scientists are doing an awful lot of damage to the world in the name of helping it. I don't mind attacking my own fraternity because I am ashamed of it," Dr. Mullis says.

He jumps to his feet to swat a yellow jacket that has infiltrated the cabin.

Given his success as an independent thinker in chemistry, Dr. Mullis's challenges to other kinds of orthodoxy are not to be lightly dismissed. But the line between fact and entertainment in Dr. Mullis's world can be hard to discern. In his book he professes to believe in astrology, to have been rescued from a fatal accident by a person traveling in an astral plane, and to have conversed with an alien disguised as a raccoon.

Asked why people should accept his views on AIDS when he has no standing as a virologist, he replies, "I don't care, I'm on my vacation life here."

Vacation life? "I have a spiritual thing in me. After lots of tough lives I got a vacation." Is he speaking of reincarnation as a metaphor or the real thing? "I believe it," he says. "If reincarnation is a useful biological idea it is certain that somewhere in the universe it will happen."

Is it not awkward to accommodate reincarnation within the theory of evolution? "I don't think DNA is the whole thing even though I invented a cool way of playing with it," Dr. Mullis declares.

The invention of PCR may well become a paradigm of scientific discovery because of its significance and because Dr. Mullis has described the eureka moment so graphically. But though the idea was central, proving it worked was also important. According to Dr. White, the friend who got Dr. Mullis a job at Cetus and oversaw part of the development of PCR, the reduction of the idea to practice was done largely by others.

"Mullis as an experimentalist is sort of hit and miss," Dr. White said. "He got a lot of data but he was having personal problems and tended to do uncontrolled experiments, so it wasn't very convincing when he did get a result."

Even after a year, Dr. Mullis had not developed definitive proof of his concept, so Dr. White then enlisted another scientist. Within a few months, Dr. Randall Saiki, a rigorous experimentalist, produced data that convinced everyone at Cetus that the process worked, Dr. White said.

Dr. Mullis believes his colleagues tried to take credit for the invention away. Dr. White denies that, saying a plan to have Dr. Mullis author the first paper describing the theory of PCR went awry because Dr. Mullis whiled away the summer creating fractal pictures on Cetus's computers instead of doing experiments.

By default, a paper by Dr. Saiki and other scientists on the applications of PCR was published first. Dr. Mullis's own paper was rejected by the journals *Nature* and *Science* on the ground that it was not new.

"I feel he has never accepted responsibility for the course of how the publications came out," Dr. White said, noting that he and colleagues attempted to let scientists know Dr. Mullis was the inventor by such means as having him describe the technique at an important conference.

Dr. White pays tribute to Dr. Mullis's fertile mind, describing how he came up with practical ways to improve PCR, such as the use of Taq polymerase, an enzyme made by bacteria that live at high temperatures. But managing his friend's creativity was not a carefree task.

"He's a hard person to know, hard not only on his spouses but on his friends," Dr. White said. "In the midst of being extremely charming, he could be extremely abusive." The two men are no longer close friends, but Dr. White, now vice president of Roche Molecular Systems in Alameda, California, owned by Roche Holding, said that Dr. Mullis was "a very unusual person, no doubt about it—I am happy I knew him."

Dr. White's version of events is supported by Dr. Paul Rabinow, an anthropologist at the University of California at Berkeley, who made a study of Cetus at the time PCR was invented. His book, *Making PCR,* was published in 1996. "Mullis is a brilliant, gifted guy who at Cetus found himself protected by a very steadfast character, Tom White," Dr. Rabinow said. "The one person who never said he wanted credit for PCR was Tom White."

As for the monetary rewards for PCR, Dr. Mullis says in his book he was "plenty wrong" in his assumption that he would be amply rewarded by Cetus. His former colleagues consider he did not do too badly. He voluntarily quit the company in 1986 at a time when no commercial value had been established for PCR, and five years before its sale to Roche for $300 million. "If the guy had been around five years later he would have been handsomely rewarded," Dr. White said. Dr. Levenson said: "Any invention you make is owned by the company. That's the deal."

Dr. Mullis has blazed through his friends' lives like a meteor, leaving so blinding a trail that few feel they see the core. "I don't know where creativity comes from," Dr. Levenson said. "He built his model of the universe to fit what he observed."

In Dr. Rabinow's view, Dr. Mullis is "a tinkerer, a bricoleur, he loves to play with things, he loves to try things out, he ignores people who say you can't do it." He adds, "He was an experimentalist not in the high scientific sense but the magician sense."

In Dr. Mullis's new profession, as author and lecturer, the magic is less evident. His book is amusingly written, but some of its viewpoints seem a little ad hoc, like a surprising attack on the Federal Reserve Board as a "tawdry sepsis."

Dr. Mullis repeats the words several times to savor their resonance. It's a good fighting phrase, but why apply it to the Federal Reserve? Dr. Mullis explains that with the Board's ability to intervene in currency markets its members have ample opportunity to profit from their inside knowledge. "If you can get around the law you do it, and Alan Greenspan is no different from Kary Mullis," he declares.

"But Kary, you're not dishonest," his wife protests. "With money I am," he says defiantly. The bottle of red wine at his side, full three hours ago, now stands empty. The yellow jackets are resuming their campaign. A mind that made a brilliant invention is wandering between sense and solipsism.

—NICHOLAS WADE
April 1998

THE SHORTCUT TO LIFE'S SECRETS

Most scientists must endure lengthy training before arriving at a position where they can make any discovery of consequence. Kary Mullis had no fondness for formal learning. After getting a Ph.D. he dropped out of science for several years. Then, while employed essentially as a technician, he made a discovery of extraordinary originality and importance, the invention of PCR.

Short for polymerase chain reaction, PCR is a method for amplifying any chosen segment of DNA on the chromosomes, the immensely long molecules in which the hereditary material is packaged. The amplification is the essential first step for almost all manipulations of DNA. Without enough material to work with, no analysis can begin.

In the days before PCR, the only method of amplifying a chosen region of DNA was to splice it into the chromosome of a bacterium, grow the bacteria until they had divided many times, and then extract the DNA, a laborious task that took several weeks. With PCR, amplified DNA can be available in a matter of hours. Moreover, PCR transfers the amplification process from the unreliable innards of bacteria to the crisp certainty of a chemical reaction in a test tube.

The idea of the technique rests on a pecularity of the DNA polymerase enzyme, which is cells' biological machine for copying DNA. The polymerase requires a short piece of double-stranded DNA at its starting point. In PCR, this short piece, called a primer, is synthesized chemically and added to a solution of the double-stranded DNA to be copied, along with samples of the four bases, or chemical letters, of which DNA is made.

The sequence, or order of units, of the primer is chosen to match the sequence at the start of the DNA strand to be copied. A second primer matches the stopping point (in fact, the starting point on the other strand: the two strands of the DNA double helix are anti-parallel).

The mixture is heated to separate the two strands of the DNA, and then cooled, which allows the primers to bind to their chosen sites on each single strand. A polymerase enzyme then starts to copy the single strands, and soon there are twice as many double-stranded DNA molecules as before. The whole mixture is heated again to separate the double strands. Most enzymes would be disrupted by the heat, but Dr. Mullis used a polymerase from a bacterium that lives at high temperatures in a deep-sea vent. The Taq polymerase, as it is known, survives and goes to work copying the DNA molecules once the mixture has cooled again and the primers have attached. At each cycle the original DNA sample is doubled, and within a few hours has increased a billionfold.

The fact that a stopping point for the polymerase is not defined doesn't seem to matter: the chains soon get chopped down to the right length. Because all components of the system are heat-stable, the whole procedure can be automated: ingredients in, flip switch, amplified DNA out.

Dr. Mullis invented this amazingly elegant procedure while trying to solve a quite different problem, that of how to replace a single base in a DNA chain. Though Nobel prizes are usually given to scientists who make big discoveries, it is the inventors of techniques who make new discoveries possible. Allowing DNA to be manipulated and analyzed chemically, instead of bacterially, was a step that changed the practice of molecular biology. Scientists could now fly instead of plod. Dr. Mullis richly deserved the prize that came his way in 1993.

—N.W.

Daniel S. Goldin

Bold Remodeler of a
Drifting Agency

Associated Press

Born in the Bronx, street smart and full of energy, Daniel S. Goldin is the kind of person who can get comfortable in a crowded restaurant of booths and counters, of gruff waitresses and $2.50 cheeseburgers. While in New York on business, he downs a burger, fries and milkshake while recounting the challenges of heading the National Aeronautics and Space Administration, the Federal Government's largest scientific agency, a faded star whose luster he is trying to restore. His dark business suit, softened by wire-rim glasses and a battered leather briefcase, somehow fits the restaurant's decor and lunch-hour din.

Suddenly his face brightens. He leans forward in a confidential manner. Did you know, he asks, his voice dead serious, about the 15th century Portuguese navigator who learned to catch prevailing winds halfway across the Atlantic toward Brazil before coming about and heading east, thus avoiding the doldrums and creating an indirect route to Africa that turned out to be much faster? That kind of breakthrough, that kind of "nonlinear thinking," he says, is essential to the rebirth of NASA.

It is the only way to do more with less, to turn bureaucrats back into innovators, to explore new worlds in an age of fiscal belt-tightening. Appointed 20 months ago in 1991, the 53-year-old Administrator, a former executive of TRW Inc. who pioneered civil spacecraft and spy satellites, began a crusade that shows no signs of losing vigor. Right or wrong, silly or sagacious, Mr. Goldin has pounded through one change after another at NASA, championing new ways of thinking, replacing the agency's top leadership, articulating a new agenda he sums up as "smaller, cheaper, faster, better."

Predictably, his changes have generated cries of alarm and howls of protest, accompanied by charges of incompetence and repeated rumors of his impending replacement. Some of the loudest criticisms have come from NASA veterans, aerospace contractors and agency patrons on Capitol Hill, all of whom have watched their influence wane. The crusade has also produced no little turmoil and turbulence in NASA's rank and file.

But, surprisingly perhaps, it is also starting to produce some results.

Mr. Goldin laid the groundwork for the recent successful repair of the Hubble Space Telescope by insisting that the astronauts redouble their practice sessions, both on Earth and in space. He has pushed through plans for a flotilla of no-frills unmanned space probes, which have ambitious goals despite their small sizes and budgets. And, in what is perhaps his greatest coup, he has saved, at least for the moment, the planned $30 billion space station at a time when Congress is eager to ax big-ticket items.

Repeating the mantra, "one world, one station," Mr. Goldin brokered a deal among Congressional barons and unruly international partners to have Russia join the building and use of the orbital outpost for astronauts. In a stroke, he helped transform the beleaguered decade-old embarrassment into a foreign-policy pillar of the Clinton Administration.

If Mr. Goldin's plans succeed, he will have given the White House a major reason to care about NASA for almost the first time since the Moon landings. "There was an event that defined the end of the Cold War—

when the Berlin Wall came down," Mr. Goldin said recently while lunching at another Manhattan restaurant, this time the comparatively upscale Century Cafe, where he looks equally at home. "There is no event that can better define the coming of the new age than we join with Russia and actually invest in technology instead of building weapons," he said. "We have to do something that has a positive benefit for the people on the planet."

Mr. Goldin's critics say his upbeat attitude is belied by actions that are arrogant, rude and often destructive, acts that have crippled the agency, perhaps for years. But some concede that even a miracle worker would be hard pressed right now because NASA, with the Cold War's demise, has lost its original reason for being.

"He's in at a bad time," said Dr. John L. McLucas, a former Air Force Secretary and former chairman of NASA's Advisory Council, a group of independent experts who review agency policy. "He hasn't done a very good job. He hasn't made that many friends. He's a hip shooter. His employees are afraid of him—and maybe for the wrong reasons, because they don't know what the hell he's going to do next. It's O.K. to have a certain amount of fear because of being held to a higher standard. But with him, people are never sure."

The friends that Mr. Goldin has made during his tenure in Washington include the most important for a NASA Administrator, those currently in the White House. The accomplishment is particularly notable since Mr. Goldin was a Bush appointee.

"He's done a great job under very difficult circumstances," Vice President Al Gore said in an interview. "He's honest. He's smart. He works hard and his heart is in the right place. I presided over his confirmation process in the Senate. I liked him from the start. I know he's met a lot of resistance at NASA and has had to do a lot of learning. But I'm for him. I think he's doing a beautiful job." The jump from Republican to Democratic camps is one of many creative leaps—some might say inconsistencies—in Mr. Goldin's life. He is a cold warrior eager to make peace with the Russians in space, a veteran of top-secret Government projects who thrives in the limelight, a small-is-beautiful type presiding over the birth of the biggest collaborative science project of all time, a blunt-talking, hard-edged New Yorker who can drift into linguistic fuzz about nonlinear thinking and managerial excellence.

He is, in some respects, a bundle of contradictions that somehow seem to work for him rather than against him.

"My judgment is that he will rank among the greatest, if not the greatest, of NASA Administrators," said John E. Pike, head of space policy for the Federation of American Scientists, a private group in Washington. "I think he's well on the way to meeting the unprecedented challenge of moving NASA from a Cold-War to a post-Cold-War world."

Daniel Saul Goldin was born in the South Bronx on July 23, 1940, the first of three children. His father, a biologist by training, worked at a post office for many years before getting a job as an elementary school teacher, eventually becoming a principal. The Goldin home was in a Jewish and Italian neighborhood across from Public School 93, where Daniel went to elementary school. His father made him feel different from the neighborhood kids.

"When everybody was outside playing baseball, he made me practice clarinet and took me to museums," Mr. Goldin recalled, praising his father as a major force in his life. The boy showed a love of things mechanical, building model airplanes, rockets and jet-powered cars.

Eventually he studied mechanical engineering at the City College of New York, working at places like Macy's to help earn his keep. In October 1957, his freshman physics teacher wrote, "Sputnik is watching you," on the blackboard as the Russians orbited the world's first artificial satellite, and Mr. Goldin instantly became a space nut. After graduating in February 1962, he applied for just one job—with NASA, newly created amid the East-West space race.

He went to NASA's Lewis Research Center in Cleveland because, Mr. Goldin said, "they were working on electric propulsion for going to Mars." His work eventually led to a bright idea, a nonlinear one. The ion engine he was working on for space propulsion could be converted to a radio transmitter powerful enough to beam television signals from a satellite to Earth, speeding them across thousands of miles of space. His advancement of that idea eventually won a United States patent and helped give birth to direct-broadcast satellites, which increasingly circle the Earth. After moving to Cleveland, Mr. Goldin married Judith Linda Kramer, the two having met in New York years earlier when they both worked at Macy's.

In 1967 Mr. Goldin was hired by TRW, a maker of military and civilian spacecraft based in Cleveland. He moved to its California divisions and stayed there for 25 years, rising through the ranks. Between 1976 and 1983 he managed several top-secret programs involving such military gear as spy satellites. Amid this highly classified work, Mr. Goldin found himself increasingly involved in what he calls executive "tea sessions" meant to hone managerial skills. Around 1980, he studied under and befriended Dr. Moshe F. Rubinstein, an engineering professor at the University of California at Los Angeles who writes books on creative problem-solving and heads seminars on the topic for business executives. "He was a shining star in that program," Dr. Rubinstein said of Mr. Goldin. "He's the kind of person who is impatient with mediocrity, with lazy brains. I have tremendous respect for him."

But some subordinates at TRW found Mr. Goldin despotic, ill-tempered and prone to grandiose plans with small hope of fulfillment, a complaint echoed later at NASA.

"I like to take people to some level of their discomfort zone," Mr. Goldin said during an interview in his NASA office. "One problem in our society is that we don't reach high enough. People are afraid of punishment and ridicule. I put stress on my staff by asking them to reach higher. It's difficult but important. It doesn't always win popularity contests for me, but lots of people come back later and say they were glad I did it."

In 1987 Mr. Goldin became general manager of TRW's space and technology group, where he suffered a major blow when a contract for a costly military satellite was canceled. "I had to lay off many people, people I had known personally," he said, clearly upset by the memory. "I got very sick" physically.

Meditating on the setback, he decided that advances in electronic miniaturization would soon allow the building of spacecraft that were smaller, lighter, cheaper and smarter than ever before, reversing a decades-long trend to bigness. Additionally, a host of small projects and contracts would make a manufacturer less vulnerable to financial disruption.

While winning contracts for the new approach, Mr. Goldin also discovered that TRW's proposal for a small, cheap satellite for studying the Earth's environment met stiff resistance at NASA, so much so that the agency ordered TRW to withdraw a proposed paper on the topic from a public meeting.

"I became very angry and I vowed I would make a change—I was furious," Mr. Goldin recalled, clenching a fist, his wrath palpable. His opportunity came when the White House seized on Mr. Goldin as a way to shake up an agency it found remarkably unresponsive. In announcing the nomination on March 11, 1992, President Bush praised Mr. Goldin as "a man of extraordinary energy and vitality."

He arrived at an agency that had been in a tailspin since the 1986 *Challenger* disaster that killed seven astronauts, including a high school teacher. Its wobbly state was driven home to Mr. Goldin two months after his nomination, when balky hardware aboard the space shuttle *Endeavour* forced three astronauts to reach out with nothing but their gloved hands to snare a wayward satellite in space. The maneuver put the shuttle and crew at some risk of injury and collision.

After that, Mr. Goldin rapidly ordered a study to see if added rehearsals and training were needed for the agency's greatest impending challenge, repair of the $1.6 billion Hubble Space Telescope. Eventually, shuttle astronauts conducted a record three preparatory space walks. The job went with surprising ease and alacrity, giving the agency its first big confidence-building accomplishment in years. President Clinton, in a telephone call to the orbiting astronauts, praised their work as "an immense boost."

On that day, December 10, Mr. Goldin credited the success to NASA's rediscovery of teamwork.

"Everybody is calling each other by their first names," he said in his office, beaming with pride for NASA's 24,000 civil servants.

His shelves are lined with pictures of his wife, Judy, and two daughters, Ariel and Laura. He boasts of being a new grandfather. A coffee mug is emblazoned with "Carpe Diem," "Seize the Day" in Latin. A small model of the American eagle sits next to a Russian bear carved out of wood. Above its head, the bear holds a winged space shuttle, the defunct Russian one, known as Buran. The model was a gift of the Russian space agency. "They're literally putting it into a park," Mr. Goldin said of the Russian spaceship, which flew only once into space and never carried aloft an astronaut crew.

Among his achievements Mr. Goldin lists the beginning of a series of small probes meant to pioneer studies of Mars, gravity, asteroids and innovative ways to make spacecraft even smaller, cheaper and smarter. He also hopes to begin work on a small probe meant to explore Pluto, the most distant and mysterious of the planets.

One place the cheaper-faster-better ethic could backfire is NASA's space shuttles, which are very complex and temperamental. Mr. Goldin is pushing to streamline their operations amid a debate over whether recent variations in engine thrust could tear the spaceships apart if the irregularities worsen. NASA officials deny any danger, saying safety is still paramount while conceding a struggle is on to cut costs wherever possible.

Perhaps the toughest job Mr. Goldin took on at NASA was wrestling with the space station. First proposed by President Ronald Reagan in 1984, the outpost had grown in size and its planned completion date had slipped even as engineers spent $9 billion just for plans amid a host of redesigns. At Mr. Clinton's request, Mr. Goldin began exploring ways to shrink the behemoth. As the work progressed, his staff kept coming back with costly models showing each new twist.

Enough, Mr. Goldin cried. He borrowed a Lego building block set from a colleague with children, using the toy to test new configurations for the outpost. He proudly displays the Lego parts, which he keeps in a cardboard box in his desk drawer. In December in Moscow, Mr. Goldin signed a formal agreement to have the Russians join the planned international space station, which, in its final form, is to orbit the Earth in the year 2001.

"This is it," he told reporters with no hint of nonlinearity. "We intend to build hardware. No more redesigns."

—WILLIAM J. BROAD
December 1993

NASA'S BUMPY RIDE

Seven years and some $100 billion after Daniel Goldin arrived at NASA eager to shake up an agency adrift in failure and mediocrity, the revolution that he engineered has lurched repeatedly from triumph to crisis and shows no signs of slowing down.

The greatest success of "faster, cheaper, better" was the $270 million Mars Pathfinder mission of 1997, which became a global sensation as its plucky little robot, Sojourner, rolled across the dry Martian soil, analyzing rocks and sending back stunning photographs.

The greatest blow came in late 1999 as two spacecraft meant to continue the Mars adventure both disappeared. The $356 million pair consisted of an orbiter to track the Martian climate and a lander with retrorockets as well as two small hitchhikers the size of cantaloupes that were to smash into the planet at 400 miles per hour to probe beneath the surface. Both types of landers were to search for signs of water ice.

The longest continuously serving head of the National Aeronautics and Space Administration, Mr. Goldin at 59 has remained zealous, even as he appointed panels to investigate the Mars woes. He said he was chastened by failure but unbowed in his tenets. "We got set back," he said in a December 1999 interview. "But we'll pick ourselves up, dust ourselves off, and we're going to rock this world. The revolution continues."

He said his initiatives had produced dozens of successes in space exploration—discovering new worlds and probing many others, from the Moon to the galactic maternity wards where stars are born. The explorations, he asserted, have also saved taxpayers at least $40 billion.

His upbeat appraisal is all the more remarkable given a rash of failures beyond the Mars setbacks. Two Earth-observing craft, Lewis and Clark, failed in orbit in 1997 and production in 1998, the victims of instrument problems, testing delays, rising costs and control-system flaws. In December 1998, the Near Earth Asteroid Rendezvous spacecraft missed its meeting with Eros, a potato-shaped rock 25 miles long. It is to try again in 2000. The telescope on the Wide-Field Infrared Explorer satellite lost its protective cover shortly after launching in March 1998, letting coolant escape into space and ruining the astronomy mission.

Despite such troubles, several analysts note that NASA under Mr. Goldin's guidance has had many more successes than failures. Winners include spacecraft that have studied the Moon, the Sun, an asteroid, the Earth's magnetosphere, interstellar clouds, interplanetary dust and distant stars. And NASA has twice refurbished the Hubble Space Telescope, which continues to push back all kinds of frontiers.

Howard E. McCurdy, a space historian at American University who is working on a book about Mr. Goldin's tenure, noted that the total cost of 16 small missions was about the same as that of the Cassini spacecraft, a $3.3 billion colossus of the old school that is heading to Saturn.

"That's why it's so attractive," McCurdy said of the Goldin revolution. "You can do 16 missions for the price of one." He added, "You lose maybe 20 percent, but you're still ahead."

In the future, Mr. Goldin's star fleet is to focus increasingly on the riddle of whether life exists beyond Earth. The two most likely sites in the solar system, Mars and Europa—an icy moon of Jupiter that may harbor a fluid ocean—are to be thoroughly explored. For Mars, inexpensive probes are to blast off in 2001, 2003, 2005, 2007 and so on. By 2008, NASA hopes to find and return to Earth for detailed analysis samples of Martian life, either fossil or extant. Even a simple microbe, scientists say, would prove for all time that life is an inherent property of matter and that the cosmos is probably aswarm with aliens.

—W.J.B.

Wallace S. Broecker

Iconoclastic Guru of the Climate Debate

If there is a great guru among the relatively small group of scientists who try to fathom the present and future behavior of the Earth's climate by studying how it has behaved in the past, it has to be 66-year-old Wallace S. Broecker: part crusty curmudgeon and blunt-spoken iconoclast, part imp and practical joker—but all intellect.

"I have a mind that leaps around a lot," says the perpetually youthful geochemist at Columbia University's Lamont-Doherty Earth Observatory, whose still-reddish hair sets off a leprechaun's twinkle. That leaping brain, he says, has led some people to accuse him of having a short attention span. But many colleagues say nimbleness of mind has enabled Wally

Broecker, more than any other person, to set the research agenda in the burningly relevant field of paleoclimatology.

His influence, they say, stems from a prodigious knowledge acquired over four decades of how the Sun, the oceans, the land, the atmosphere, ice and vegetation work together to shape the Earth's climate; an ability to fashion that knowledge into a coherent picture and a quickness to discern how emerging bits of information and insight fit into or alter the picture.

A good scientist, Dr. Broecker once said, ought to be working on 27 things at once, pushing each idea as far as it can go, dropping it when progress stalls, then coming back to it later. "No scientist can do that," says Dr. Richard Alley of Pennsylvania State University, regarded as no slouch in the field himself. "But Wally can."

Dr. Broecker's office on the Lamont-Doherty campus just west of the Hudson River at Palisades, New York, is stuffed with bric-a-brac that suggests a mind not only nimble but also playful and slightly off center. Among the items are a Mickey Mouse hat, a dried piranha, a Dolly Parton poster and the brightly painted head of a Neanderthal man on a mannequin wearing a dress and a badge that says "Scrooge Was Right." "I like odd things like that," Dr. Broecker said.

But there is no computer to be seen. A computer, he explained, "short-circuits" one's thinking about how the Earth-ocean-atmosphere system really works. So, although he sometimes relies on other people's computer analyses of parts of a problem, he mostly puzzles out his ideas about once-and-future climates with a pencil, on plain white paper.

The product of Dr. Broecker's thinking that has propelled him into public view of late is his statement that "the climate system is an angry beast and we are poking it with sticks." The reference, of course, is to the human impact on the climate system, chiefly through emissions of heat-trapping greenhouse gases like carbon dioxide, a product of burning fossil fuels like coal and oil.

Disturb the climate system enough, Dr. Broecker argues, and it may react with unpleasant surprises. "The Earth's climate is very volatile; it can do some weird things," he says. Among the weirdest are sharp, rapid warmings and coolings that can transform the global climate in a human lifetime or even within a decade.

Dr. Broecker was one of the earliest to advance the view that such changes can occur, in the mid-1980's, and he has been its foremost proponent ever since. Lately, chemical clues of past changes, extracted from ocean and lake sediment and ice sheets, have proliferated, bolstering his pioneering conclusions.

Dr. Broecker has linked these abrupt changes in past climates to the on-again, off-again action of vast, deep ocean currents, mightier than a hundred Amazon Rivers, that transport great amounts of heat around the planet, shaping the climate of much of the globe. This "great conveyor belt," as he calls it, is perhaps the idea with which he is most closely associated.

The genesis of his interest in the idea, back in 1984, was a colleague's revelation that ice age climates appeared to shift sharply back and forth between two stable states. What could cause this? The next day, using his knowledge of the deep ocean current, he sat down and calculated that it delivered fully 30 percent as much heat to the North Atlantic as comes from the Sun. "Whoo-ee!" he thought. Turn the conveyor off, and the impact on climate would be sudden and huge.

Today, he makes no bones about his belief that if the world warms as much as computer models predict, just a few degrees, it could set off a chain of events that would abruptly slow or stop the conveyor, with disruptive results for much of the world. The most disruptive, he believes, would be brief but large "flickers" in global temperature as the climate readjusts itself not smoothly, but in fits and starts.

In a paper he has been writing lately, he postulates the somewhat surprising idea that cutting off the conveyor late in the next century, as computer models suggest might happen, would make Europe somewhat cooler than it is now, while at the same time partly offsetting whatever global warming takes place in much of the rest of the world over the next century.

The bad part "would be the on-and-off flickers, because they'd be unpredictable," he says with a flat Chicago accent. In the past, he pointed out, such flickers have produced drastic changes in temperature within periods as short as five years. If this happened in the future, he says, it could mean disaster for world agriculture at a time when the Earth's human population might have doubled.

He is quick to note that this grim possibility hinges on a lot of things, not least the accuracy of the computer models that predict both how warm the world will get in the decades ahead and how the warmth might affect the workings of the conveyor. How good are the models? "Who knows?" he says.

Dr. Broecker ponders these problems starting in the morning at his home study in Closter, New Jersey, after the cats have been fed, the paper has been read and he has chatted a bit with his wife, Grace, with whom he has raised six children.

Among his intellectual attributes, acquaintances say, are a skepticism that leads him to home in on exceptions to the rule, which jars him out of intellectual complacency and stimulates further thinking, and an extroverted personality as volatile as the climate system he studies. He is known for blunt honesty, a self-confessed volcanic temper and a reluctance to suffer sloppy thinkers gladly, balanced by a sense of fun that has made him an incorrigible practical joker.

A few years ago, when cultural and scientific openings to China were relatively new for Americans, a top Chinese scientific official visited Dr. George Kukla, Dr. Broecker's close friend and a colleague at Lamont-Doherty. As told by Dr. Kukla, the Chinese guest had asked especially to meet Dr. Broecker. Late that evening, Dr. Kukla got into his station wagon to drive the visitor back to his hotel. He stepped on the gas and the engine revved but the car did not move; it had been put up on blocks. From the bushes came muffled laughter. Out walked Dr. Broecker and another friend and colleague, Dr. Dorothy Peteet.

"May I introduce you to Professor Broecker?" Dr. Kukla snapped in an angry voice. Embarrassed, Dr. Broecker tried to recover by asking if they made jokes like that in China. "Yes," said the visitor, "but only children." Dr. Broecker still chuckles at the memory.

"It was like that constantly," said Dr. James White, a geochemist at the University of Colorado, one of many former Broecker students who have fanned out into the paleoclimatology community over the years. At the same time, said Dr. White, Dr. Broecker "could just go absolutely red in the face, livid, mad at me," adding, "But 10 minutes later all is fine, and he never got mad on personal things; it was always on the science, if I was doing something dumb." The Broecker attitude was, "You've got to respect the science."

Dr. Broecker says: "There's something sacred about science. I get my delight out of finding the truth but I also don't like to see it monkeyed with."

Once, confronted by a Federal bureacrat's demands that he be specific about what a particular piece of research would produce, he told the official that he produced ideas and that ideas could not be predicted in advance. Later he tried to have the man dismissed, and when a higher official vacillated, Dr. Broecker slammed down a copy of *Profiles in Courage* on the man's table and stalked out. "Wally plays to win," says a colleague.

Dr. Broecker, whose father ran a gas station in Chicago, grew up in a fundamentalist Christian home and attended Wheaton College, a conservative Christian institution, before transferring to Columbia in 1952. He now describes himself as areligious, but the relationship between science and a fundamentalist upbringing has been a frequent source of discussion with Dr. Peteet, who had similar experiences growing up in Georgia. In Dr. Broecker's case, Dr. Peteet believes, the Christian insistence on honesty took firm root and infuses his scientific work. But his skepticism, she says, is probably traceable to rebellion against the dogma and blind faith he encountered as a youth.

This may be, she says, why both she and Dr. Broecker tend to focus not on overall patterns of climatic functioning that constitute conventional wisdom, but rather on "things that don't fit." For instance, she said, there was a time when scientists at Lamont believed that periodic changes in the Earth's position relative to the Sun, then a particularly popular idea, explained the ebb and flow of ice ages. But Dr. Broecker focused on a prominent exception to the rule: a brief, sudden plunge into near-glacial conditions, called the Younger Dryas period, that occurred about 12,000 years ago, after the last ice age had already ended.

Digging for answers to the unexplainable, said Dr. Peteet, is "what makes somebody good," adding, "He gets excited about it." (The changes in the relative positions of the Earth and Sun still are considered the basic trigger for ice ages, but it turned out that the most likely explanation for the Younger Dryas was a cessation of the ocean heat conveyor.)

The most productive thinking, says Dr. Broecker, happens when "you realize that something you've held as a truth is wrong, and therefore everything that depends on it unravels, so then you're in confusion and you start to rethink the problem." He learned early never to "consider any thought you've had as being sacred, because that's the worst thing you can get into."

While Dr. Broecker is viewed mainly as an idea man, he scarcely functions in a vacuum devoid of concrete data. He began his career as director of the radiocarbon dating laboratory at Lamont, and his textbook, *Tracers in the Sea,* is still standard reading for scientists who use chemical clues to investigate climate. The shells of foraminifera, tiny ocean animals whose bodies carry the chemical traces of past climatic change, are Dr. Broecker's stock in trade. A co-worker has framed some pieces of popcorn that uncannily resemble the tiny animals, and placed them on Dr. Broecker's wall with the label, "Wally's Bugs."

Dr. Broecker says he sees one of his functions as that of discerning and "keeping track of bottom lines" in the welter of new knowledge about virtually every aspect of the Earth-atmosphere-ocean system. "It's hard to find something he hasn't touched," says Dr. Alley. This is clearly reflected, Dr. Alley says, in indices that keep track of which scientists are most frequently cited by others: "If you open up to the Broecker page, it's usually the Broecker pages; it's just phenomenal the impact he has."

He has taken a few of what some people might see as improbable fliers. Once, he and a colleague investigated the feasibility of neutralizing global warming by using airplanes to spread droplets of sulfur dioxide through the stratosphere. These would reflect the Sun's heat, cooling the planet. They calculated that it could be done by using 700 Boeing 747's to release the droplets around the clock, at a cost of about $40 billion a year. The general response, he says, was, "Well, Wally, this is interesting, but the world is not ready for this kind of stuff." He believes, however, that the strategy is "an intelligent insurance policy" if the climate changes catastrophically and people someday get desperate.

Will people deliberately manipulate the Earth's climate on a large scale in the future? "Much as I dislike it," he says, "you can see that the temptation is going to be very high and the capability to do it will become very high, and so whether we'll resist, I don't know."

Dr. Broecker could retire, but he has no plans to do so and is not resting on his laurels. "No one lives on their past successes," he says. "It isn't very satisfying. You live on what you're doing this year, this month. My great joy in life comes in figuring something out. I figure something out about every six months or so, and I write about it and encourage research on it, and that's the joy of life."

—WILLIAM K. STEVENS
March 1998

A SUDDEN CHANGE IN THE WEATHER

Scientists once assumed that when big changes in climate take place, they do so slowly and gradually. But in the late 1990's, they were finding more and more evidence that this relatively comforting version of reality is not always right—and that the opposite may, in fact, be fairly common in the Earth's history.

In the earlier version, experts thought that the climate system responded to what they call "forcings"—for instance, rising atmospheric concentrations of heat-trapping gases like carbon dioxide, or stronger solar radiation—much as a stereo set does: turn up the volume and the sound gradually gets louder. In the new view, it often responds more like a thermostat: nothing happens until a certain threshold is reached, and then there is a sudden switch in state.

Some of the most intriguing evidence comes from the last 10,000 years and the centuries immediately preceding, when the world was emerging from its most recent ice age.

By examining a variety of chemical tracers extracted from glacial ice, scientists have found that when the world began its final ascent out of the last ice age more than 11,000 years ago, temperatures in Greenland spiked upward by some 15 degrees Fahrenheit within only two decades. Another Greenland ice core showed that the switch from ice-age conditions to something like today's came in two quick temperature jumps, each of nearly 10 degrees and each lasting less than a decade, within a 40-year transition period.

These sharp jumps were bigger than that for the globe as a whole, but they apparently reverberated into more southerly climes as well. Studies of lake pollen in the New York City region revealed that during the glacial-to-interglacial transition, the average annual temperature shot up by 5 to 7 degrees within no more than 50 years.

To get an idea of what this means, it helps to realize that the average global temperature is only 5 to 9 degrees warmer now than in the depths of the Ice Age 18,000 to 20,000 years ago.

Evidence has also turned up of more recent abrupt shifts. For instance, one study found that roughly every 1,500 years since the end of the last ice age, the climate of the subtropical Atlantic has flipped to a sharply colder state within 50 to 100 years or less. This translated into severely curtailed rains in Africa, so severe that they might have wreaked disaster on some early civilizations.

In the more remote past, experts have discovered that about 55 million years ago, after the super-warm age of the dinosaurs ended but before today's continuing cycle of recurrent ice ages began about 2.5 million years ago, the global climate warmed abruptly. In the process, many forms of marine life were wiped out, but the warmer climate enabled the newly established class of mammals to expand and diversify rapidly. Apparently, the sudden upward spike in temperature developed when a more gradual warming reached a certain threshold, setting off a complex chain of feedback effects that amplified the warm-up.

Could that happen again as a result of today's global warming? Is there a threshold of warming at which the climate system could suddenly shift to an entirely different state? If so, how far away is that threshold?

No answers are forthcoming so far, but the questions are among the most serious being investigated by climate scientists in a greenhouse world.

—W.K.S.

Michael E. DeBakey

Dr. DeBakey at 90: Stringent Standards and a Steady Hand

Sam C. Pierson Jr. for The New York Times

Dr. Michael E. DeBakey moved like a dancer in a pas de deux, anticipating his surgical partner's every move during a quadruple bypass operation on a woman at the Methodist Hospital in Houston. As Dr. DeBakey deftly manipulated scalpels, scissors and forceps for more than three hours, his hands never quivered.

With the rib cage cut open to expose the chest cavity, Dr. DeBakey gently pulled up the thin tissue surrounding the heart. His partner, Dr. George P. Noon, sliced into the tissue. The diseased coronary artery came clearly into view. The surgeons opened the artery and sewed part of a vein taken from the leg into place as a graft. They did this four times, in one place using an artery as a graft, to create new channels for blood to flow around clogged sections of the diseased coronary arteries.

What made it an extraordinary operation is that Dr. DeBakey, the pioneering heart surgeon, was just weeks from his 90th birthday, on September 7. With bushy eyebrows, long dark hair that is combed back on a balding scalp, and a tinge of gray on his sideburns, Dr. DeBakey looks 20 years younger. He is celebrating his 50th year at Baylor College of Medicine, and only a disability will make him retire, he says.

Baylor has no age limit on its surgeons, Dr. DeBakey said. He operates less often these days, not because he is not up to the task, but because he does not have time to do the requisite follow-up after every operation. He is too busy pursuing his other interests as international medical consultant and teacher.

"I have been away more than I have been here recently," he said. "Operations tie you down."

The same night he performed the operation, he attended a couple's golden wedding anniversary party. "You haven't changed a bit," friends told him. "Aren't you nice to say that," a smiling Dr. DeBakey replied. "I have been blessed so far."

A recent checkup found Dr. DeBakey in the physical shape of a man much younger, with normal heart function, cholesterol and other blood tests, he said. He has been a hospital patient twice. Once was for surgery for a near-fatal bleeding ulcer in 1984. The other was for smoke inhalation he suffered in rescuing his infant daughter, Olga, after a Christmas tree caught fire in his home in 1978. "I just barely made it out," he says.

As a child growing up in Lake Charles, Louisiana, Michael DeBakey's usual fare was fish and fresh vegetables from the family garden. Now, he says, he consumes whatever he wants but is a light eater and often skips lunch because of work. He said he has never deliberately exercised, rarely takes vacations and last visited his 1,000-acre ranch in the nearby Texas hill country two years ago.

Trim at 160 pounds and 5 feet 11 inches, Dr. DeBakey can fit into his Army uniform from World War II. In those years, he helped develop the mobile military army surgical hospitals, now popularly known as MASH units, and systematic follow-up of veterans with certain conditions.

As a civilian, Dr. DeBakey went on to pioneer many heart and blood vessel operations, perform at least 65,000 surgeries, teach and build Baylor

College of Medicine into a top-flight institution, serving 10 years as its president (from 1969 to 1979) and then 17 years as its chancellor (from 1979 to 1996).

Though now so firmly rooted in Texas that he wears gleaming white cowboy-style boots in the hospital, Dr. DeBakey has stamped out a glamorous career as an international medical statesman. He experienced a taste of life abroad while doing surgical training in Europe in 1935 and 1936 and speaks Arabic, French and German. Even now, he has set a two-week schedule that includes visits to Aspen, Colorado, Brussels, Istanbul, Stockholm, New York and Washington.

Outside the United States, Dr. DeBakey is probably best known in Russia, which he has visited at least 30 times to perform operations and to work on health programs. It was 88-year-old Dr. DeBakey, not a younger American surgeon, whom President Boris N. Yeltsin's doctors summoned to Moscow for consultation for Mr. Yeltsin's quintuple bypass operation in 1996. In a foreword to the Russian edition of Dr. DeBakey's 1997 book *The New Living Heart,* written with Antonio Gotto (Adams), Mr. Yeltsin described Dr. DeBakey as "a magician of the heart" and "a man with a gift for performing miracles."

Around Baylor, Dr. DeBakey's inexhaustible energy has led to the nickname "The Texas Tornado," and his high standards and sometimes harsh demeanor have intimidated some staff members. When young doctors failed to meet his expectations, Dr. DeBakey said, he would rebuke them: "You have four fingers and a thumb like I do. Why can't you use them the way I do?"

Dr. Noon, a former student and trainee who began working with Dr. DeBakey 30 years ago and who now takes many of his surgical cases, said that over the years "there were many Dr. DeBakeys—the chancellor, chairman of the department of surgery, the teacher, the national and international leader." He often held the positions simultaneously. "He was good at getting everything organized, but delays and disruptions made him mad," Dr. Noon said. "Now his schedule is not as demanding and he is less high strung."

Asked about his reputed temper, Dr. DeBakey said in his Louisiana drawl, "Maybe I should be more compassionate with people who do not think as fast as I can." But he added, "I have little tolerance for incompetence, sloppy thinking and laziness."

A lengthy visit in late August attests that Dr. DeBakey still hates to waste a minute. Two days before performing the bypass operation, he went straight to the office upon returning from a long weekend trip to Azerbaijan, where he had advised President Heydar Aliyev on the creation of a new 100-bed heart center.

Dr. DeBakey lives a five-minute ride from the hospital in the same house he has lived in for 50 years. The day he performed the operation, Dr. DeBakey awoke at 5 A.M., had fruit juice and yogurt for breakfast, read a newspaper and worked on a paper analyzing long-term findings on thousands of operations he had performed. Before 8 A.M., he left for the hospital, where he made a videotape promoting an experimental device his team is developing to assist failing hearts.

Dr. DeBakey heads the search committee for a top position at the Baylor-affiliated Veterans Administration Hospital, and after returning from the operating room, he interviewed a candidate. Later, he reviewed the special heart X-rays known as angiograms of three patients and advised them on the need for surgery. Dr. DeBakey left the hospital about 6:30 P.M., as he does on most days.

On this evening, his family was away. His second wife, Katrin, an actress and artist he met as a widower at a party at Frank Sinatra's home, was visiting in her native Germany. Olga, their daughter, is spending her junior year in college studying in Lebanon. (He has four sons from his first marriage; none are doctors.) Dr. DeBakey spent nearly three hours at the anniversary party, drove home in his Toyota Land Cruiser, and went to bed about 11.

On other days, Dr. DeBakey worked on additional projects chiefly related to heart disease, which is the leading cause of death in developed countries. He talked about women and heart disease at a luncheon sponsored by the American Heart Association; advised doctors on plans to develop new cardiovascular centers in China, Lebanon, Turkey and Russia; worked on arrangements for the first implantations of the heart-assist device in patients in Germany, Sweden and Switzerland; and consulted through the Internet.

His path to medical celebrity was set, indirectly, by his parents, who left Lebanon to escape religious persecution and settled in Lake Charles, in the French-speaking Cajun country. They owned rice farms, real estate and drugstores, which became hangouts for doctors. "I became enamored of them," Dr. DeBakey said.

In the late 1930's, Dr. DeBakey and his mentor at Tulane University, Dr. Alton Ochsner, both non-smokers, became two of the first scientists to link the early rise in lung cancer to cigarette smoking. At the time, other medical leaders attributed the cancer increase to factors like the 1918 influenza epidemic or inhalation of poisonous gases in World War I.

Dr. DeBakey's mother taught him to sew. In the early 1950's, he drew on that know-how to make grafts from Dacron and use them to repair aneurysms, or balloonings, in all parts of the aorta. Such grafts are now part of standard treatment.

"I really opened that field," Dr. DeBakey said.

Inventions like the roller pump, used in heart-lung machines, have become standard in medicine. These days, Dr. DeBakey heads the medical advisory board of Micromed Technology Inc. of Woodlands, Texas, which is refining a ventricular assist device.

Peers acknowledge Dr. DeBakey's many talents, as well as his ample pride. "No question, he has a pretty big ego," said Dr. Claude J. Lenfant, the head of the National Heart, Lung and Blood Institute, a Federal agency in Bethesda, Maryland. But, Dr. Lenfant added, "his contributions have been enormous and he will leave an amazing legacy." Unlike many other famous surgeons, "Dr. DeBakey has exported his know-how to the world," said Dr. Lenfant, who first met Dr. DeBakey when he demonstrated open-heart operations as a visiting professor at Dr. Lenfant's medical school in France in 1953.

Dr. DeBakey has a knack for telling a good story and an appetite for publicity. A framed copy of Dr. DeBakey on the cover of *Time* magazine on May 28, 1965, is among hundreds of mementos in his office.

When he first arrived in Houston, his rising profile did not make him popular with doctors in the local medical society. "They did everything to denounce me because I was a threat to them," Dr. DeBakey said. "Everything I've done has been resisted," he said, citing such problems as revising the standard of surgical training in Houston; overcoming the resistance of doctors and nurses in starting the first intensive care units at Baylor, and admitting black patients to the hospital.

Now, however, he is treated with the deference of a living legend. Around Houston, Dr. DeBakey is recognized instantly. At the hospital, employees ask if they can have their pictures taken with him, and he complies. Even during the quadruple bypass, a young surgeon stationed himself behind Dr. DeBakey so he would be in a photograph of the master performing the operation.

After the operation, Dr. DeBakey took the elevator and inserted a key that allowed it to climb nonstop from the third-floor operating room to his ninth-floor office. The key was a compromise; a benefactor wanted to build a private elevator for Dr. DeBakey, who rejected that idea and settled for the key.

Dr. DeBakey freely talks about the rich and famous patients he has treated. Sometimes it is to instruct, as in his American Heart Association talk to women when he described the operation he performed on the actress Marlene Dietrich to relieve blockage of arteries in her legs. He also sprinkles his conversations with stories about meeting Queen Elizabeth and royalty from Belgium, and the generosity of patients whom he proudly said he recruited as donors. Dr. DeBakey said he told wealthy patients like Stavros Niarchos and Aristotle Onassis, who asked about his fees, that he would not send them a bill. Instead, bringing up his foundation, "I said, I would like for you to consider making a contribution." The foundation now contributes $2 million a year to Baylor's research program, Dr. DeBakey said.

Dr. DeBakey's persuasiveness has shown up on the political front as well. After World War II, he played a major role in getting a decrepit library in Washington run by the Army replaced with the National Library of Medicine at the National Institutes of Health in Bethesda. Dr. DeBakey thought the library should be near Government researchers in Bethesda, but at first, he said, James A. Shannon, the NIH director, did not agree. Dr. DeBakey convinced Dr. Shannon otherwise by asking him how he started his pioneering kidney research at New York University. By reading in the library, Dr. Shannon replied. But the American Medical Association wanted the library built next to its headquarters in Chicago, and with little interest in the issue, Sam Rayburn of Texas, the House Speaker, held up the bill backing the Bethesda site, which was sponsored by Senator Lister Hill of Alabama. Dr. DeBakey, though, had once operated on the husband of the secretary of the national Democratic Party, Dorothy Vredenberg, and he asked her for help. The next day she told Dr. DeBakey that Mr. Rayburn would let the bill through. The Bethesda building now houses the world's premier medical library.

Medical wings at Baylor and around the world bear Dr. DeBakey's name. Other buildings do, too, including a Houston high school he helped start 25 years ago for students, mostly minorities, seeking health care careers.

Dr. DeBakey is also proud of the 1,000 American and foreign surgeons he has trained. Many of those surgeons are working in top positions, including Dr. Renat S. Akchurin, who performed the quintuple bypass on Mr. Yeltsin. Yet, Dr. DeBakey said: "I'm shocked to hear some are retired. Where did the time go?"

—LAWRENCE K. ALTMAN, M.D.
September 1998

THE DOCTOR AS RESEARCHER

The path to medical advances often goes full circle: observations made at a patient's bedside stimulate research in the laboratory that produces new therapies that are brought back to the bedside.

Much of our knowledge about human digestion owes to the observations that an Army doctor, William Beaumont, made on the bullet-torn abdomen of Alexis St. Martin. After Dr. Beaumont treated St. Martin for gunshot wounds in 1825, the wounds healed but left a hole extending from the skin to the stomach. Realizing he had a unique opportunity to peer into a human stomach to record how it reacted to various conditions, Dr. Beaumont conducted 238 tests over the next eight years and made fundamental discoveries about human digestion.

In 1929, a German intern, Dr. Werner Forssmann, wondered if patient survival would improve if he found a faster way to deliver drugs to the heart. Breaking a taboo that held that any doctor who touched the living human heart would lose the esteem of his colleagues, the intern inserted a thin tube into a vein in his own elbow crease and threaded it into his heart. The technique, cardiac catheterization, is now standard in intensive care, and it paved the way for bypass operations and other heart surgery and the modern treatment of heart attacks.

Research for Dr. Beaumont and Dr. Forssmann was an opportunistic, part-time effort. Their successes inspired other doctors to pursue clinical investigation as a career. Dr. Michael DeBakey, for example, used the sewing techniques he learned from his mother and the catheterization method developed by Dr. Forssmann to pioneer operations to repair damaged arteries and hearts. One was surgery to repair a bulge in the aorta, the body's main artery, before it fatally burst. Another was to bypass clogged arteries in the pelvis and thigh to restore blood flow and relieve leg pain.

Dr. DeBakey, like scientists before him, went where the funding was. Louis Pasteur, for example, competed for grants from industry to make his early discoveries. When Dr. DeBakey trained as a surgeon in New Orleans, the limited available funding came largely from private philanthropists. There were no formal courses in research; young investigators learned as apprentices.

Then, after World War II, American taxpayers made increasingly large investments in medical research. The goal was to improve the nation's health, although there was no explicit national goal like putting a man on the Moon.

With greater public support, a bright and dedicated physician could spend a year or two training in a laboratory and then combine practice and research in a university hospital. The number of medical schools and affiliated hospitals expanded. Patients and doctors teamed up to improve the accuracy of diagnoses and better understand how diseases progress naturally to cause damage.

Instead of studying individual cases, doctors organized studies of larger groups of patients, expanded them to include a number of hospitals, and analyzed the findings with improved statistical methods. New drugs, devices, other therapies and preventions came rapidly. As death rates from tuberculosis and other infections dropped sharply, doctors turned their attention to chronic diseases like those affecting the heart and joints.

However, as funds for research came to dominate the budgets of medical schools, some overly enthusiastic clinical investigators violated ethical standards. Federal health officials ordered the creation of institutional review boards, whose approval would be required for any experiment on a human.

In addition, the great post-war advances in molecular biology posed profound challenges to clinical investigators. As this laboratory field thrived, resources shifted from clinical research conducted by medical doctors to laboratory studies at the cellular and molecular levels performed by Ph.D.'s.

In the late 1970's, leaders in American medicine began calling clinical researchers "an endangered species." Young doctors seeking careers in clinical research faced the obstacles of heavy debts from medical school tuition and increased competition from Ph.D. investigators. More recently, new threats have arisen from managed care organizations that frown on supporting research.

Still, as long as taxpayers support doctors who want to explore the medical puzzles they encounter in their practice, the clinical researcher is likely to survive.

—L.K.A.

Meave Epps Leakey

The New Leader of a Fossil-Hunting Dynasty

Reuters

Two families have met in Africa in the 20th century, one seeking the origins of the other, in the course of which both have emerged from obscurity. They are the Leakey family, the now celebrated dynasty of fossil hunters, and the early hominids, the ancestors and close relatives of modern humans about whom almost nothing was known before the Leakeys began finding their bones, stone tools and footprints.

The Leakey and hominid family trees have become so entwined that it seemed appropriate, perhaps even inevitable, that the most recent announcement of a major fossil find should herald not only the discovery of a new hominid species, but also the accession of a new Leakey to leadership of the family's continuing search for human origins.

In true dynastic fashion, the new leader is Meave Epps Leakey, daughter-in-law and wife of those who established and sustained the Leakey reputation in paleoanthropology. But no one is gainsaying her qualifications. If anything, years of hard-earned experience in the field and laboratory, along with respected academic credentials, make her better prepared than any previous family member to carry on the Leakey tradition.

So far, the burden of being a Leakey and keeper of the family flame seems to rest easily on Meave Leakey's shoulders. At 53, she is slender and tanned from years under the African Sun, her light brown hair beginning to gray. She moves with indoor grace through a round of lectures and receptions in her honor. But her hands are of the outdoors, the long, strong fingers seeming to be ready to brush dirt off another jawbone or deftly fit the pieces of a shattered cranium. Her manner is reserved, though anything but defensive or unfriendly.

In an interview recently at Rutgers University in New Brunswick, New Jersey, where she delivered a lecture, Dr. Leakey spoke of her strategy of exploring deeper in time than any Leakey before her. Not that she seems to feel a need to prove anything, except the nature of hominids more than four million years ago as they began walking on their hind limbs soon after their fateful split from the ape lineage.

"I had a choice of concentrating on gaps in the time period already being worked or going to earlier sediments," Dr. Leakey said. "We found fossils in the earlier sediments clearly worth a look. It was so much more exciting. We now have a chance of finding the earliest hominids or ancestors of both apes and hominids."

And if the vaunted Leakey luck holds, she just might succeed.

Questions about being a Leakey by marriage brought a smile telegraphing reticence. "It's never a dull moment, ever," she said. "They have so many interests beyond paleontology. It's quite challenging."

The dynastic transition is a big step for Meave Leakey and possibly for paleoanthropology. She becomes one of the most visible scientists in a discipline that has been a male domain; when she was starting in the 1960's, women were regularly denied places on expeditions. And people who have worked with her describe Dr. Leakey as a self-effacing person and cautious scientist given to understatement, both rare traits in a highly competitive field that often encourages flamboyant and contentious behavior.

"She conveys an integrity and honesty that one does not experience everyday with people in this field," said Dr. John W. K. Harris, an archaeologist at Rutgers, who did field work with Dr. Leakey in the 1970's.

The dynasty's founder, Louis S. B. Leakey, operated in the flamboyant mode but made discoveries that established East Africa as the ancestral grounds for many human forerunners. Son of British missionaries in Kenya, Louis Leakey was a largely self-taught paleontologist who began exploring Africa's fossil past in the 1920's. Upon his death in 1972, his widow, Mary, a trained archaeologist, carried on and expanded the work with steady success, notably at the Olduvai and Laetoli sites in Tanzania. She is now 82, retired but still smoking little cigars.

After Louis died and Mary eventually slowed down, their son Richard stepped into the leadership role. At first, he had been reluctant to enter the fossil-hunting game. When he did, he gained an early reputation for brashness, perhaps driving himself to eclipse his father's fame and do it before a kidney illness might cut short his life. With his health restored through a kidney transplant, Richard mellowed and won praise for leading expeditions with outstanding geologists and anatomists at his side. He hit pay dirt with discoveries of several early *Homo* specimens around Lake Turkana in northern Kenya.

"He's matured into a real mensch," said Dr. Adrienne Zihlman, an anthropologist at the University of California at Santa Cruz.

When the London-born Meave Epps graduated in zoology from the University of North Wales, she responded to a newspaper advertisement for a job in Kenya and was hired in 1965 by Louis Leakey to care for monkeys at a primate research center. This enabled her to complete research for her doctorate. Later, she worked with Richard Leakey, dissecting monkeys. They fell in love, and in 1969, she joined him for field work at Lake Turkana. They were soon married.

In *Ancestral Passions: The Leakey Family and the Quest for Humankind's Beginnings,* published by Simon & Schuster, Virginia Morell writes, "At times, Richard's life has borne an uncanny resemblance to his father's, and his affair with Meave in some ways mirrored Louis's with Mary."

Both father and son were married when they fell in love with their protégées and eventually left their wives. Their second wives became accomplished collaborators, each bringing to the work an academic standing superior to their husbands', though it was much later before the wives could emerge from the shadows of their more outgoing husbands.

Meave Leakey recalled in the interview: "With Louis and Mary, Mary was the person who did the detailed work. Louis was the organizer, the effervescent fund-raiser, the public person."

Has this also been the case with Richard and Meave? "Yes," she conceded, without elaboration. Other scientists remembered visits by the two in which Richard did all the talking, while Meave hung quietly in the background, and also a movie of Leakey explorations that showed Meave at a camp but never named or identified her. She is naturally shy, Dr. Zihlman said, and appeared to be more comfortable staying out of the spotlight.

Dr. Leakey began to come into her own in 1989, assuming most of the responsibilities of directing field work while her husband became increasingly active in conservation as head of the Kenya Wildlife Service. Then, in an airplane crash two years ago, Richard lost both legs below the knees.

"It's your show now," Dr. Leakey recalled her husband saying to her at the time.

Like others in paleoanthropology today, Dr. Leakey takes a more sophisticated approach to research in human origins than used to be practiced. She works closely with a team of other scientists to reconstruct the environment in which the early hominids lived, looking for clues to explain their evolution from apelike creatures to upright-walking, tool-making and large-brained creatures.

"Louis used to find a fossil, describe it and move on to a new species," she said of her father-in-law. "Now people have any number of techniques for analyzing fossils and ancient environments. We can ask a whole lot of questions that we could not ask before. We are focusing more on asking those questions."

The foremost questions concern the development of the larger, more humanlike brain and the even earlier transition to upright walking, or bipedalism. Since sediments bearing fossils between one million and three million years old had already been well explored, revealing outlines of the evolution of the large-brained *Homo* genus, Dr. Leakey decided to look for older sediments where she just might find fossils of some of the first bipedal hominids. She found them 30 miles southwest of Lake Turkana at a site called Kanapoi.

In September 1995, Dr. Leakey reported the discovery of 4.1-million-year-old fossils of a previously unknown hominid species, which has been named *Australopithecus anamensis.* The leg and arm bones provide the earliest direct and unambiguous evidence for bipedalism.

Her find filled a wide gap in the fossil record. Earlier in the year in Ethiopia, Dr. Tim D. White, a paleontologist at the University of California at Berkeley, had uncovered fossils of a 4.4-million-year-old hominid, named *Ardipithecus ramidus;* whether it was bipedal has not been established. Before then, there was little evidence for hominids older than 3.6 million years. These were creatures called *Australopithecus afarensis,* the most famous skeleton of which, Lucy, was found by Dr. Donald C. Johanson of the Institute for Human Origins in Berkeley.

And paleontologists have only begun to explore this critical period of time, soon after the hominid lineage is thought to have diverged from the ape line.

"You never get a novel adaptation like bipedalism without a radiation of species, a number of different experiments in this adaptation," she said. "I'm sure when more evidence is in, we'll have more species, and the more species you get, the more complex will be the picture of origins."

In recent decades, the Leakeys have had running personal and professional feuds with Dr. White and particularly Dr. Johanson. But Dr. White has now invited Dr. Leakey to examine his new specimens, and she said she hoped that relations with Dr. Johanson could be repaired. "It's no good working in isolation," she said. "Whatever you think of your colleagues, you should talk."

Richard Leakey remains keenly interested in paleoanthropology, raising money for the projects and often visiting his wife's dig sites on weekends. But it is the present, not the past, that is now his consuming interest. In May he helped found a new political party in opposition to Kenya's autocratic president, Daniel Arap Moi, a step filled with risk. Richard and several of his party supporters were assaulted recently by young thugs, severely beaten and whipped.

This puts Dr. Leakey, as head of paleontology at the National Museum of Kenya, in a difficult position. "I work for the Government, so I stay well clear of politics," was all she would say on the matter.

Richard has moments of regret at having to give up active fossil hunting. "Nothing can match the thrill of finding a fossil hominid," Dr. Leakey said. "He must miss that. I know he misses it."

The Leakey role in paleoanthropology is likely to continue beyond Meave. Their 23-year-old daughter, Louise, has a degree in geology and biology, solid background for a career hunting fossils. She has worked with her mother at Turkana and impressed colleagues.

Will she carry on the family tradition?

"She doesn't want me to answer that," Dr. Leakey said. "She's making up her own mind."

—JOHN NOBLE WILFORD
November 1995

FRESH NEWS FROM THE PAST

Hardly a year goes by without more news out of Africa of fossil discoveries bearing on early human evolution. Each report is a cryptic message from the dim past and species long extinct, a message hinting at revelation but beyond ready decipherment.

Among the most promising recent discoveries was a 2.3-million-year-old upper jaw from the badlands of northern Ethiopia. The jaw is the most convincing and earliest definitively dated fossil of the genus *Homo,* to which living humans belong. It extended the established age of the direct family line by 400,000 years, closer to the time of the first evidence of tool making and to environmental upheavals that may have been decisive in human evolution. As such, the fossil provides a rare glimpse into what has been a kind of dark age of evolutionary change, the period between three million and two million years ago.

In another major find, scientists in South Africa pieced together what they describe as the best-preserved and most nearly complete skull and skeleton yet seen of any very early member of the human family. These were the remains of a four-foot-tall individual who lived between 3.2 million and 3.6 million years ago. The specimen is older and much more complete than the famous "Lucy" skeleton found in Ethiopia in 1974.

This discovery promises to yield important insights into a critical phase of human evolution, a time of transition in hominid locomotion. Ample evidence has already shown that these apelike australopithecines were walking upright by this time. But were they exclusively upright walkers? The first traces of this skeleton suggested that this species was also capable of grasping and climbing like a chimpanzee. These upright walkers on the ground might have fled back into the trees for safety or foraging.

As these discoveries were being announced, Meave Leakey continued the slow, gritty excavations into the even earlier period of evolutionary transition that has been the focus of her recent work: the time immediately after hominids split from the ape lineage. If she and her team, including Alan Walker, have made any new striking finds in Kenya, they are keeping the news to themselves until they are sure.

But Dr. Leakey's *Australopithecus anamensis,* discovered in 1995 and dated at 4.2 million years old, stands as the earliest species known to walk on two legs. Tim White has yet to report further details of *Ardipithecus ramidus,* a more ape-like being that lived 4.4 million years ago in Ethiopia; the claim that it also walked upright is based on only indirect evidence.

As a mark of their endorsement, other paleoanthropologists have placed *A. anamensis* at the root of their most current versions of the human family tree. Dr. Leakey's discovery gave her peers one more reason to reconsider their earlier, simpler ideas about what that family tree looks like. The tree resembles less a tall poplar with straight trunk, representing a single, linear transformation of one species into another all the way to modern humans, than a bush with the many branches of a meandering, multifaceted evolution.

Reflecting the new understanding, Dr. Ian Tattersall of the American Museum of Natural History in New York City wrote in the January 2000 issue of *Scientific American* that the evolutionary history of hominids "is marked by diversity rather than by linear progression."

Since no one believes that all of the prehuman species have been found, paleoanthropologists like Dr. Leakey keep combing the hills and valleys of Africa for more fossils. Her daughter, Louise, is often at her side. Her husband, Richard, is still more engaged in politics and conservation, leaving fossils to his wife and daughter. Mary Leakey, matriarch of the Leakey dynasty, died in December 1996 at 83.

—J.N.W.

A Scientific Passion for Wolves

David Binder/The New York Times

The paw prints in the reddish sand of a forest road were only vaguely defined by the time Jim Hammill saw them, softened by an overnight rainfall: the tracks of a timber wolf.

Usually gladdened by any sign of a wolf, Mr. Hammill was disheartened because these prints were made by only a single animal, and they were more than a mile from where he had set a line of five traps along a wolf path. He was searching for members of the Nordic Pack, a wolf family that has been ranging around 130 square miles in north Dickinson County in Michigan for several years.

Wolves migrating from northern Minnesota through Wisconsin recently re-established themselves here in the Upper Peninsula of Michigan more than three decades after they were nearly extirpated—a period during which the only wolves around were wanderers from other regions.

"For the last couple of days, I haven't been able to find hide nor hair of the pack," Mr. Hammill, a wildlife biologist with the State Department of Natural Resources, said on a gloomy morning before the onset of winter.

Canis lupus lycaon, the Eastern timber wolf, was indigenous to Michigan beginning about 10,000 years ago as the ice age glaciers retreated in North America. Now, with a verified population of 30 on the heavily forested Upper Peninsula, wolves stand a precarious chance of becoming permanent inhabitants again. They resumed breeding in Michigan in 1991, for the first time since 1954. "We can't say the wolf is here to stay," Mr. Hammill said. "It's too early."

His task is to try to preserve the Upper Peninsula as a wolf habitat, which means maintaining adequate prey, chiefly white-tailed deer, while guarding the wolves from their chief predator: man. Tall, rugged and soft-spoken, Mr. Hammill, who is 44, is one of the younger people among those Dr. L. David Mech, the wolf expert from Minnesota, counts as 30 or 40 specialists around the world. They share a passion for the animal he calls "the shyest of all creatures."

The descendant of Cornish miners who came to the iron and copper country here a century ago, Mr. Hammill has been curious about wolves since his boyhood in the town of Norway, but he did not see one until 1974. Nine years later he saw clear wolf tracks on a snowmobile trail in north Dickinson County and felt a thrill. "When wolves are there, it's a magical place again," he said.

In July 1992, Mr. Hammill snared a timber wolf, the first under the state's recovery program for wolves. The program was then in its infancy. He listed the wolf as 01. "I didn't want to give him a name," he said. Fitted with a $200 radio collar while briefly tranquilized, 01 was monitored along with other members of Nordic Pack for seven months until he died on the shore of a little lake.

Mr. Hammill spotted the lifeless animal on February 7, almost depilated by mange. "It died of exposure, hypothermia," he said. "It was 21 degrees below zero." He had inoculated the wolf against mange and other diseases when it was trapped, but it was obviously reinfected, perhaps from contact with coyotes. "When I saw him lying there in the snow with wolf tracks all around, my heart sank," he recalled. "That night I sat up in bed thinking about him and shed a few tears, not for him so much, but for the whole wolf-management program. I felt it was set back five years."

Now, half a year later, he was trying to trap another wolf to attach a radio collar. The technique of tracking wolves with radio telemetry was developed by Dr. Mech in the Superior National Forest in Minnesota in 1968. Since then it has been successfully used in Canada, Alaska and Western Europe.

"It gives an idea of the movement of the entire pack," Mr. Hammill said, "as well as the longevity of an individual and, if you follow them to the den, how many pups are born, how many survive the first, most critical winter."

Once the range of a wolf pack has been established—five packs with three or more members have been verified on the Upper Peninsula—Mr. Hammill and his associates are in a position to make their habitats a little more tenable. "We can close roads or decide not to build roads," he said, "closing areas that people don't need to access, usually areas that are being timbered." Among Michigan's blessings is the fact that the state is one of the largest landholders in the country, with a total of 4.5 million acres. Mr. Hammill keeps an eye on state forest lands totaling 750,000 acres in seven counties.

Until recently, most deer hunters regarded wolves as rivals, even as foes. That is reflected in some of Mr. Hammill's experiences. In 1974, Michigan tried to move four wolves to Marquette County from Minnesota, but within eight months all had been killed—three by hunters and trappers.

In 1985, a wolf that had been given a radio collar in Wisconsin was shot at close range in Iron County on the second day of deer hunting season, Mr. Hammill said. An informant reported to Mr. Hammill's office in Crystal Falls that the man who shot the wolf had chopped off the radio collar and thrown it into a river, while keeping the wolf cadaver in his woodshed. Mr. Hammill retrieved collar and carcass, "but law officers couldn't prove the man killed the wolf," he said.

There has been a vast change in attitudes since 1974, when F. E. Noble, a militant hunter in Baraga County north of here, organized a Wolf Hunters Association with the slogan "Preserve Our Deer—Shoot a Wolf" and offering a $100 reward for any wolf. This, despite a 1965 state law protecting wolves and a law placing wolves on the endangered species list in 1971.

Today, a great majority of Michigan hunters have become pro-wolf, in part because of the educational efforts of Mr. Hammill and his colleagues. A 1990 study showed that 64 percent of Upper Peninsula residents responding to a survey supported restoration of wolves to the region, while 76 percent of the hunters and trappers approved.

In 1992, the Department of Natural Resources organized a "Wolf Awareness Week" in the hope, as Mr. Hammill put it, that "the long chain of anti-wolf sentiment can be broken" and followed up last year with Wolf Forums around the state, where specialists discussed wolves' importance to the environment. They explained that wolves are needed to cull herds of deer and moose of the sick and the old and for other contributions to the balance of wildlife in the north woods.

In deer season, Mr. Hammill is on the road meeting with hunters in the field as the advocate of the wolf, distributing fact sheets and forms for reporting observation of a wolf. He also gives about 20 lectures a year.

He receives calls from a network of about a dozen wolf spotters, most of them hunters. They respect the biologist partly because he has hunted a lot himself—deer, ruffed grouse and snowshoe hare. They also enjoy ribbing him. "You should see the wolf I just shot!" bellowed Charlie Hinman, a retired state police officer, as he entered a cafe in which Mr. Hammill was eating a sandwich. Then, looking at Mr. Hammill, he said, "Oh, I didn't know you were here!"

To watch Mr. Hammill mixing with local residents is to see an ambassador at work. He joshes easily with them. But it is clear they understand the seriousness of his purpose as he explains why he did not seek prosecution of a dairy farmer who shot two wolves that killed two of his cows, or a coyote trapper who caught and killed a wolf because it had mange. "We didn't prosecute because we knew that other trappers and farmers wouldn't cooperate with us," he said.

The wolf represents only a small portion of his wildlife management responsibilities, which include deer, bears, pine martens, fishers, ruffed grouse, eagles and ospreys. But as "the most endangered species in Michigan" it is the most challenging because wolves are so elusive.

In September, for instance, he had four days between the end of a lengthy bear-dog training period and the opening of Michigan's bear-hunting season to trap a wolf. "I don't want to get a dog in a trap," he said. But with his radio-collared wolf gone, it was impossible to pin down the Nordic Pack. He took a chance and placed the traps on the sides of a forest road heavily traveled by wolves earlier in the year.

The specially built traps respond to pressure of eight pounds or more and have offset teeth that are easy on an animal's leg. They are covered with a sprinkling of dirt and baited sometimes with a few drops of female wolf urine, more often with an odoriferous concoction Mr. Hammill calls "Fascination," consisting of ground-up animal glands, oil of Venice and asafetida, a bad-smelling gum resin. For luck he tosses an equally noisome chunk of decomposing venison into the woods nearby. He wears rubber boots and gloves boiled beforehand to rid them of human scent for these operations.

The frustrating result from four days of trapping: one very angry bobcat weighing almost 40 pounds. Mr. Hammill tranquilized it with a hypodermic attached to a jab stick and freed it from the trap. If it had been a wolf he would have taken blood samples, inoculated it against mange, heartworm and parvovirus, an intestinal disease that can kill young canines, and attached a radio collar.

A day after he had pulled up his traps to make way for the bear hunters, Mr. Hammill's wolf hot line rang. Charles Dillon, a logger on the south margin of Marquette County, reported that his brother had spotted "four wolves, maybe five" where he was cutting timber. "That's the Nordic Pack," Mr. Hammill said. "The week is ending with good news."

The rest of the autumn and the onset of winter were not hospitable to Mr. Hammill's trapping attempts: rain, then snow, then below-zero temperatures. But the deeper snows of January and February are good for tracking wolves and he, with three other trackers, will soon be out by snowmobile and snowshoe in the wolf ranges, counting paw prints across the Upper Peninsula. "You try to decipher how many animals are there and what home range they are using," he said.

—DAVID BINDER
February 1994

PUBLIC HEARS CALL OF THE WILD

Fifty-seven timber wolves were counted on Michigan's Upper Peninsula in 1994, 80 the following year and in 1999, at least 174—a remarkable recovery by a still endangered species.

There might have been two more, but they were shot by hunters, said Jim Hammill, the wolf specialist with the state's Department of Natural Resources.

He was not distraught by these losses because in both cases the miscreant hunters were identified through tips "from the public." The first, a resident of

Iron County in his 30's, was successfully prosecuted under state law prohibiting killing endangered species, fined $2,500 and sentenced to 60 days in jail. "It sounds stiff," Mr. Hammill said, "but it was not so hard as if the feds had hit him." The second, who had boasted about his kill in local taverns, fled the state when he learned the law was after him.

With wolves, "people are aware that they are endangered species and they are not willing to let people kill that resource and get away with it," Mr. Hammill remarked. "The No. 1 factor for the increase is public support. People have been very supportive."

The Upper Peninsula wolves, he said, are also helped by "a very good food base—white tail deer and beaver." Wolves depend on beaver for sustenance especially in the fall and spring, and deer provide the diet all winter. Moose, which were reintroduced to the Upper Peninsula in 1985, make an occasional meal. "We know of at least two cases where wolves have killed moose, but deer are much easier for them," he said.

Support for the timber wolf has been engendered in part by Mr. Hammill's efforts as a public speaker. "I average speaking to about 2,000 people a year," he said, "in Michigan, Wisconsin, Minnesota, even Montana." Like other wildlife biologists in the upper Midwest, he has also been sought out by foreign scientists interested in wolf recovery or wolf reintroduction, among them Luigi Boitani of Italy and Ian Redman of Great Britain.

What would be the ideal wolf population for the sparsely populated Upper Peninsula, which comprises 16,500 square miles? "We know biologically that the population could stabilize at about 1,000," he said. "But I think the social capacity is substantially below that. There are some places where people are tired of having wolves near their cattle. There have been four fatal depredations. Right now the owners get 77 percent of market value for a cow and we're working on a supplemental to that."

"We could have people turn around," Mr. Hammill said. But he remains optimistic. "So far there has been no uprising against wolves."

He is hopeful, too, that wolves will someday find their way back to lower Michigan. "One winter they will make it across the ice bridge at Macinac" Strait, a distance of a few miles, he said, noting that wolves had done just that from Canada, crossing the ice from Sault Ste. Marie, Ontario, to the Upper Peninsula.

He continues to trap wolves and attach radio collars, but no longer alone, as he did in the early 1990's. Now he has a team of six part- or full-time aides to keep up with the packs, and to do "whatever it takes to get the animal delisted as an endangered species."

—D.B.

Master of Molecule Manipulation Works on the Wild Side

Susan Spann for The New York Times

An atom can be here, there and somewhere else all at once; it can be going both this way and that way; it can be rattling with internal energy and quiet. Such are the quirky freedoms of the quantum microworld that do not apply to automobiles, apples or people. But if there is an exception to this rule, it would be the mind of Dr. Steven Chu, the master of manipulating atoms and molecules who shared the 1997 Nobel Prize in physics with two other researchers.

Dr. Chu, a professor at Stanford University, ranges so widely among topics of interest to him that any attempt to keep up leaves a blurry succession of images, like a slide show inadequately describing one's dream vacation. Here, he is clutching a Reuben sandwich while filling a steno pad with sketches of an experiment to stretch DNA molecules like knotted rubber bands. Here, he is in a cab, alternately shouting instructions to the driver and quietly rationalizing his mega-celebrity status in Taiwan. This one shows him explaining the cosmological theory of multiple universes to the Dalai Lama. And finally, here, in his lab, Dr. Chu is pointing through the window of a vacuum chamber to a hovering blob of atoms that are trapped and supercooled by the forces of laser light—the achievement for which he won the 1997 Nobel Prize.

On a recent day in his Stanford office, the 50-year-old, aggressively trim Dr. Chu, wearing khaki trousers, a pullover shirt and scruffy tennis shoes, sat on a puffy couch and held court for his students and research associates. He bounced between taking phone calls and bantering about projects that combine conceptual simplicity with a faint air of unbelievability. In one experiment, the drop of a "fountain" of atoms is clocked to measure gravity so precisely that the technique could be used like a petroleum dowser to find hidden pockets of oil. In others, a single DNA molecule marked with fluorescent dyes is dragged about as a video camera mounted on a microscope captures the reptilian motions.

"I'm drawn towards the wilder, science-fiction-type stuff," said Dr. Chu, who grew up in Garden City, Long Island. "It's much more fun for me to learn completely new things than it is to sort of ice the cake," he said. "Well, that's part of the reason one is a scientist. It's not to just crawl into a little corner of some field and go to all of the meetings that all 50 of you in the world go to. 'Oh, this person twiddled that and look what happened.' After a while it becomes a little rococo, if you know what I mean."

That adventurousness is reflected in Dr. Chu's recent moves from his original field of atomic physics into polymer science, chemistry and biology. But unlike the average scientific dilettante, Dr. Chu sees to it that this crosscutting work always bears the mark of his experimental dazzle and intuition for pay dirt. "What Steve brings to science is an incredible technical virtuosity—a sort of technical showmanship," said Dr. David Pritchard, a physicist at the Massachusetts Institute of Technology in Cambridge. "There is this terrific scientific taste," added Daniel Kleppner, also of MIT "He has the vision to see new areas emerging. He's a damn good experimenter, but he's also highly creative."

Those who work most closely with Dr. Chu say he also has the incalculably useful ability to see the complex interactions of the microworld directly, rather than relying on the equations that thicken the chapters of physics textbooks.

"He doesn't have to resort to those formulas," said Cheng Chin, one of Dr. Chu's doctoral students. "He has atoms in his mind."

Steven Chu emerged from a cauldron of academic overachievement. His father, who emigrated from mainland China to study chemical engineering at MIT, retired as a professor at Brooklyn Polytechnic and is a member of the Academica Sinica, Taiwan's most distinguished scholarly society. His mother studied economics at MIT and a prestigious university in China. Two brothers and four cousins eventually generated three M.D.'s, four Ph.D.'s and a law degree. And that doesn't count the aunts and uncles.

"I was to become the academic black sheep," Dr. Chu wrote in a biographical sketch for the Nobel Foundation. There is still an edge to his voice when he recalls that for years in his father's eyes, he never lived up to the dizzying scholarly achievements of his older brother, Gilbert Chu, an associate professor at Stanford School of Medicine.

With what he described as a "relatively lackluster" record at Garden City High, Dr. Chu was not accepted into Ivy League colleges. Instead, he attended the University of Rochester, where his intellectual appetite exploded, leading to undergraduate degrees in physics and math and a doctorate in physics at the University of California at Berkeley.

An encounter with the late Richard Feynman's *Lectures in Physics,* a comprehensive and idiosyncratic introduction to basic physics, ignited an interest in theoretical physics. Not until later, well into graduate school, did he start fiddling with equipment in physics laboratories at night. It was like Nat King Cole finding his voice: so long, jazz piano. "I said, 'Well, it's ridiculous, sneaking in and making play experiments, and I'm ignoring my theoretical problem,' " Dr. Chu explained. Experiments became the day job.

He quickly began a long-term relationship with the powerful light source called the laser. His method was to juice up the inadequate equipment he could buy off the shelf, which let him stay ahead of the beat scientifically. That improvisational approach served him well when he moved to the old AT&T Bell Laboratories, the famed center of technical and scientific innovation in New Jersey.

"He would come into my office with these great ideas, all excited, and scribble stuff up on my board," said Dr. Douglas Osheroff, who shared the previous Nobel Prize in physics and is now a professor at Stanford. (The university has raked in five physics Nobels in the past 10 years.)

As to the cross-disciplinary flair Dr. Chu would develop, Dr. Osheroff said, "that was a very standard approach at Bell Laboratories and almost part of the culture there."

It was at Bell Labs that Dr. Chu met Dr. Arthur Ashkin, who was trapping microscopic particles with laser light and who dreamed of trapping atoms. Dr. Ashkin and his colleagues were the first to demonstrate that particles would "stick" to the laser beam for nearly the same reasons that dust sticks to a glass rod that has been charged positively by rubbing it with cat's fur. Even though a dust mote has no net charge, its negative charges—electrons—crowd over to the side closest to the rod, being pulled toward it like vacationers waving at the rails of a docking ship. That leaves a positive excess on the opposite side of the mote, but because the negative charges are closer to the rod, their attraction wins out and the mote sticks. In the Bell Labs apparatus, the dancing electric fields of laser light take the place of those static electrical charges, attracting a particle to the most intense parts of a laser beam. Instruments that move particles around with lasers have come to be called "laser tweezers."

After adding this technique to his repertoire, Dr. Chu expanded on it by "gluing" tiny polystyrene beads, which he could grab and move with laser tweezers, onto strands of fluorescent DNA. By stretching the DNA and dragging it through a soup of other DNA molecules, he and colleagues at Stanford later helped confirm theories developed by Dr. Pierre-Gilles de Gennes, the Nobel Prize winner at the École Supérieure de Physique et de Chimie Industrielles in Paris. Dr. de Gennes had predicted that any such long molecules, called polymers, should "reptate," or change directions like a snake, whose entire body follows the same path without cutting corners.

"We've had theories of that kind of thing for years," said Bruno Zimm, a chemistry professor at the University of California, San Diego. "Actually seeing it is something else."

Among other things, reptation helps explain the toy-store mystery of why Silly Putty bounces and squishes. If a child hurls it against a wall, the impact comes faster than the polymers in the putty can move along their tortuous paths, so the putty acts like a solid and bounces. But a slow squish gives the polymers time to slither around. Academics call such materials "viscoelastic."

Still more recently, using a different method for manipulating the polymers, Dr. Chu and his colleagues, Dr. Thomas Perkins and Dr. Douglas Smith, took virus DNA that was in a naturally coiled state and stretched it.

They found that identical polymers beginning from the same state can uncoil in different ways, getting hung up in strange knots and bends—behavior Dr. de Gennes called "molecular individualism."

This effect could be important in everything from the extrusion of plastics to the flow of oil in pipelines, where squirming, free-floating polymers reduce the drag when they are added to the oil—improving the flow for reasons that are still poorly understood. And, noting that with every unraveling of the virus DNA polymers they used "you have the same virus and the same genes and very, very different outcomes," he said with a laugh, "So maybe you can take that as a moral in life—that it's neither nature nor nurture."

"Polymer physicists were goo-goo, ga-ga over it," Dr. Chu said of the DNA work.

But the loudest applause came when Dr. Chu used laser light to chill atoms, so that they—like the beads—could be trapped. At room temperature this would be a hopeless task, because atoms are moving at thousands of miles per hour.

In 1985, Dr. Chu and his colleagues at Bell Labs cooled a cloud of atoms to 240 millionths of a degree above absolute zero with what he called "optical molasses." They did it by bathing the atoms in laser beams fired together from all directions, which created a kind of drag that slowed the atoms to inches per second, corresponding to the low temperature. That achievement finally allowed atoms to be laser-trapped.

"When I first heard about it, I said, 'That's incredible,'" recalled Dr. Kleppner of MIT. "And the next thing was, 'But of course; why not do that?'" The molasses analogy was a brilliant way of seeing through the immense technical challenges posed by the optical cooler, he said.

Dr. Chu shared the Nobel Prize with Dr. William Phillips of the National Institute of Standards and Technology and Dr. Claude Cohen-Tannoudji of the École Normale Supérieure in Paris. All three have made enormous contributions to the science of atom cooling and trapping. The traps have become standard equipment for creating the ultraprecise timepieces called atomic clocks. The techniques are also often a starting point for outlandish physics experiments, such as when scientists at the University of Colorado at Boulder created a new state of ultracold matter, a Bose-Einstein condensate, two years ago.

Like a juggler whose hands become ever more difficult to follow over the years, Dr. Chu has continued playing tricks with trapped atoms. In one experiment, for example, he and Dr. Mark Kasevich of Yale University used laser pulses to put atoms into "split personality" states predicted by quantum theory. Both states feel the tug of gravity as they move on slightly different trajectories. The degree to which those states are compatible when the split atoms recombine gives a measure of gravity's strength to about a part in a billion. That is fine enough to detect underground geological features, like oil pockets.

Dr. Chu has also continued the push into biology, partly in collaboration with Jim Spudich, chairman of the biochemistry department at Stanford. This aspect of Dr. Chu's work has become so well known that he was almost lured away by another university's fetching offer to co-direct a new cross-disciplinary institute. "I was tempted," he said.

He turned down the offer because of the difficult decisions the move would have entailed for his two sons, ages 13 and 16. Dr. Chu is divorced and shares in their care.

"Fundamentally, I'm happy here," he said in his office. "I'm very much against people trying to leverage themselves and their institutions. I don't want to see academics go into a state of free agency."

As high as his star has risen in the United States, it is nothing compared to his profile in Taiwan. His friend Yuan Lee—yet another Nobel Prize–winning physicist—now heads the Academica Sinica, of which Dr. Chu is a member. The combination of Dr. Chu's ties to Taiwan, his particular brand of charisma and a society that still lionizes scholars has led to hordes of cameras and reporters' shadowing him in public.

"One time he came back and said: 'Taiwan is crazy for me,'" Mr. Chin recalled. A Taiwanese professor joked that Dr. Chu's popularity was second only to Leonardo DiCaprio's for a time. "DiCaprio is first by a whole lot," insisted Dr. Chu, who said that he speaks only basic Chinese with a heavy American accent.

Even the Dalai Lama found an opportunity to talk physics with him a few years ago. "He was obsessed with finding out what physics could say about time," Dr. Chu said. "I'm surmising why," he explained. "If you're a Buddhist, you believe in reincarnation, and there's this conservation-of-souls problem," referring to the question of where new souls come from as the number of living beings in the world keeps expanding.

In his moments with the Dalai Lama, the exiled leader of Tibet, Dr. Chu outlined the speculative theory that the universe is just one of many that are appearing and disappearing like bubbles in a champagne glass. (The theory is notably championed by another Stanford physicist, Dr.

Andrei Linde.) That way, as vast amounts of time passed and whole universes became inhospitable to life, souls might flit from one universe to another and keep the total number constant, Dr. Chu suggested.

The Dalai Lama's reaction?

"He just listened," said Dr. Chu. "Juuuuust listened."

—JAMES GLANZ
June 1998

THE GENIUS AND THE CRACKPOT

Physicists, if they are successful enough to penetrate the public consciousness, learn that with fame come fans with—to put it mildly—eccentric ideas.

These cosmic theorizers, who might be defined as obsessed, confused, intellectually outmatched or just plain weird proponents of demonstrably incorrect ideas, maintain a voluminous correspondence with noted physicists and turn up at their talks and speeches with the reliability of paparazzi chasing princesses and pop artists.

"I get about a grand unified theory a month," said Dr. Steven Chu as he hurried toward an escalator after a talk. He had just disentangled himself from an argumentative man in a sport jacket who was carrying copies of several century-old papers by Albert Einstein.

Sure enough, the topic on the man's mind, as he pointed vigorously at passages in the papers, was his theory of how all of physics could be knitted together with one or two revolutionary ideas.

What may be more surprising is that Dr. Chu, a Stanford physicist and Nobel laureate, did not simply brush off someone from whom he clearly had so little to learn. Dr. Chu debated a few points with him, wished him well and gently made a getaway.

Scientists, even those with Dr. Chu's high-powered credentials, display a willingness to engage the woolliest of theorists more often than might be believed. The reasons vary, ranging from fear of insulting someone who does not seem quite stable to tolerance of a scientist who, outside of a single, inexplicable obsession, does solid work. The result is that certified geniuses and unmistakable crackpots frequently come in contact.

Einstein was no exception. Once, in the 1940's, he met for four hours with Wilhelm Reich, who believed that a box called an Orgone Accumulator could concentrate an energy of which the entire universe was supposedly composed, and that he thought could cure human maladies like cancer and impotence.

"When I told him, in concluding, that people considered me mad, his reply was 'I can believe that,' " Dr. Reich wrote in his journal. Reich apparently believed Einstein was sympathizing with him.

Explaining the origin and contents of the entire universe is irresistible to armchair theorists in general, said Dr. Tereasa Brainerd, an astronomer at Boston University whose research has been covered in newspapers. "They are always wrong, for the simple reason that the authors have not understood the basic physics that is involved," Dr. Brainerd said.

But Dr. Brainerd said she did occasionally respond to letters, "with the idea of free public education in mind." Her approach is to try to correct a few fallacies and keep the writers interested in science.

A common theme is that their theories anticipated a discovery that has just been reported in the news, said Dr. Alexei Filippenko, an astronomer at the University of California at Berkeley. Dr. Filippenko has a thick sheaf of letters from people who believe they anticipated the discovery by him and other astronomers that the universe is filled with a strange energy that is boosting its expansion.

Wild theorizing is not limited to people with no science background, said Dr. Benjamin Bederson, a physicist at New York University. "Saddest of crackpot theories," said Dr. Bederson, "come from superannuated, formerly fine scientists who late in their careers get bored doing bread-and-butter stuff." A classic example, he said, was the renowned physicist Dr. Julian Schwinger, who in old age believed he had found a scientific explanation for cold fusion, a scheme for producing energy that was soon shown to have no basis in reality.

Cranks also turn up in less rarefied areas of research. Dr. Donald Moyer, a former physicist who is now a patent agent in Chicago, said his most frightening experience was with a man who claimed that he had invented bulletproof paint. Sitting in Dr. Moyer's office and clutching a paper bag, the man said ominously that the paint was also resistant to intercontinental ballistic missiles and suggested that Dr. Moyer would want to arrange a test.

"Finally," Dr. Moyer said, "he reached into the paper bag and pulled out an 8-by-10 plaque with a portrait of Dwight Eisenhower."

To top off his pitch, the man explained that the wondrous paint was made from recycled garbage.

The fear that such people can inspire often leads to some nimble circumlocutions on the parts of legitimate scientists, Dr. Moyer said. Once, as he was discussing crackpot theorizing with a fellow physicist in his office, his colleague took out a file marked "public relations" that was filled with letters on off-the-wall theories. When Dr. Moyer asked why in the world the folder was so labeled, his colleague explained that the writers sometimes turned up in his office, "and they get really upset if you take out a folder marked 'crackpots.'"

The best strategy of all for dealing with cranks may be to tie them up with their own ideas, said Dr. Chu. Riding the escalator to his next talk, Dr. Chu hesitated not at all when a reporter asked for advice on how to respond to a paranoid person who believed that the government was controlling his mind with invisible lasers, and who wanted to know how to escape them.

"Tell him to wear aluminum-foil hats," Dr. Chu said.

—J.G.

Bruce N. Ames

Strong Views on Origins of Cancer

Michael Jones for The New York Times

When it comes to controversy, Dr. Bruce N. Ames, though not physically intimidating at 5 feet 9 inches tall, is no shrinking violet. The man who invented the leading laboratory test to screen chemicals for their ability to damage genes has years of solid science and the accolades of countless colleagues behind him when he makes such provocative and socially unpopular statements as these:

"I think pesticides lower the cancer rate."

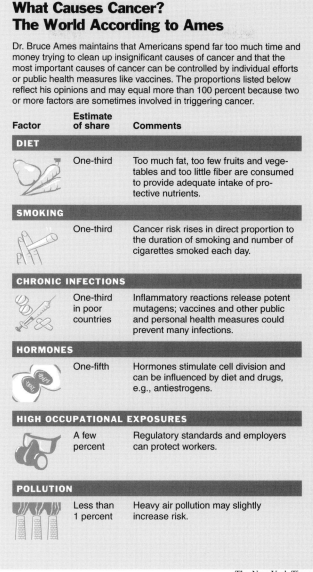

What Causes Cancer?
The World According to Ames

Dr. Bruce Ames maintains that Americans spend far too much time and money trying to clean up insignificant causes of cancer and that the most important causes of cancer can be controlled by individual efforts or public health measures like vaccines. The proportions listed below reflect his opinions and may equal more than 100 percent because two or more factors are sometimes involved in triggering cancer.

Factor	Estimate of share	Comments
DIET	One-third	Too much fat, too few fruits and vegetables and too little fiber are consumed to provide adequate intake of protective nutrients.
SMOKING	One-third	Cancer risk rises in direct proportion to the duration of smoking and number of cigarettes smoked each day.
CHRONIC INFECTIONS	One-third in poor countries	Inflammatory reactions release potent mutagens; vaccines and other public and personal health measures could prevent many infections.
HORMONES	One-fifth	Hormones stimulate cell division and can be influenced by diet and drugs, e.g., antiestrogens.
HIGH OCCUPATIONAL EXPOSURES	A few percent	Regulatory standards and employers can protect workers.
POLLUTION	Less than 1 percent	Heavy air pollution may slightly increase risk.

The New York Times

"Pollution seems to me to be mostly a red herring as a cause of cancer."

"Environmentalists are forever issuing scare reports based on very shallow science."

"Standard animal cancer tests done with high doses are practically useless for predicting a chemical's risk to humans."

"Nearly all the polluted wells in the U.S. seem less of a hazard than chlorinated tap water."

"99.9 percent of the toxic chemicals we're exposed to are completely natural—you consume about 50 toxic chemicals whenever you eat a plant."

"Elimination of cancer is not in the cards, even if we get rid of every external factor."

"Nearly half of all natural chemicals tested, like half of synthetic chemicals, are carcinogenic in rodents when given at high doses."

"We're shooting ourselves in the foot with environmental regulations that cost over 2 percent of the GNP, much of it to regulate trivia."

Coming as they do from a highly respected scientist who does not do consulting work for industry, such remarks are especially irritating to those who believe that modern industry has touched off an epidemic of cancer and birth defects by contaminating the air, water, soil and food with toxic chemicals.

Dr. Ames, a biochemist and molecular biologist at the University of California at Berkeley, where he directs the National Institute of Environmental Health Sciences Center, is a member of the National Academy of Sciences and is the recipient of an outstanding investigator grant from the National Cancer Institute and of many highly prestigious awards for excellence in research. His hundreds of technical publications, many in an arena rife with public, political and scientific controversy and punctuated with passionate emotions, have made him one of the two dozen most often cited scientists in the world.

Now 65 years old, Dr. Ames feels he is racing against the clock to "solve aging and/or cancer before my neurons decay." Not that his colleagues have noticed any dulling of his tack-sharp scientific mind, his rapid-fire ripostes and his widely admired ability to unify disparate areas of research into a common, logical construct of how cancer starts, flourishes and eventually consumes its host.

Even his detractors consider Dr. Ames to be a brilliant scientist who moves with remarkable ease from pure biochemistry into public health. Dr. Malcolm Pike, an epidemiologist at the University of Southern California in Los Angeles, applauds his "enormous ability to see the big picture" and his willingness to be "a protagonist who is prepared to get up and argue some very unpopular things." Dr. Walter C. Willet, an epidemiologist at the Harvard School of Public Health, calls him "one of the most innovative thinkers in the world of science, able to bridge the gap between laboratory science and human disease."

But others, like Dr. David Rall, former director of the National Institute of Environmental Health Sciences, believe his generalizations are based on incomplete data, since "most of the chemicals we're exposed to haven't been adequately tested for carcinogenicity."

As Dr. Ames sees it, "Much of cancer is built in; a good part of it is due to aging."

"If you plot cancer versus age," he said, "you'll see that in 60 million years, evolution took us from short-lived creatures like rats, where a third have cancer by the age of 3, to long-lived creatures like humans, where a third have cancer by the age of 80. There's very little human cancer until after the age of 30; then it increases sharply with age."

Why, he wondered, should that be? He found several explanations. First, cancer risk is linked to metabolic rate: the faster the rate at which the body burns fuel, the greater the likelihood of developing cancer. Rats, for example, have a metabolic rate eight times as fast as humans.

"What in metabolism is doing us in?" Dr. Ames asked, not waiting for an answer. "Oxidation. We burn fat and carbohydrates in the presence of oxygen to produce energy and water. But en route, we make superoxide, hydrogen peroxide and hydroxyl radicals, the same mutagenic free radicals that are produced by radiation. In each human cell the DNA is hit about 10,000 times a day by mutagenic oxidants, and in each rat cell about 100,000 times a day. You'd think that exercise would be bad, since it raises the metabolic rate, uses oxygen and increases the production of free radicals. But as you become adapted to exercise, the body's antioxidant defenses go up and the risk of heart disease goes down, so there's a net benefit."

While there is no way to stop the body's production of free radicals, he explained, human cells are replete with enzymes that block their damaging effects. These innate antioxidant defenses, he said, are bolstered by antioxidant nutrients like vitamin C, vitamin E and carotenoids in fruits and vegetables, which can suppress all stages of the cancer process.

"Diet is at least as important as smoking as a cause of cancer," he said in a tone reminiscent of parents coaxing children to eat their broccoli. "If you don't eat your vegetables, you're irradiating yourself in a sense. Gladys Block has shown that the rate for practically every type of cancer is doubled among people who don't eat fruits and vegetables. Probably every vitamin in foods is part of some defense system." Dr. Block is an expert on cancer and nutrition at Berkeley.

Dr. Ames considers pesticides an anticancer weapon because their use increases the yield of fruits and vegetables and lowers their cost, enabling more people to consume foods that appear to protect against cancer.

Dr. Block has found that even with help from pesticides, only 9 percent of Americans eat the recommended minimum of a total of five servings of fruits and vegetables daily. This dietary failing, along with a menu high in fat and low in fiber, accounts for at least a third of cancers in industrialized nations, Dr. Ames and others have concluded from analyses of cancer rates and dietary habits in many populations.

A slightly built, bookish man, Dr. Ames is a native of New York who graduated from the Bronx High School of Science and Cornell University. He says his findings about the origins of cancer and bodily defenses against it have strongly influenced his living habits.

"I never smoked, and when I married my Italian wife 34 years ago, I switched my diet from Jewish cooking to Italian cooking and never looked back," he said. "I eat lots of fruits and vegetables and fish, but meat in moderation and few processed foods. We use mainly olive oil, and each day I take 250 milligrams of vitamin C, 400 international units of vitamin E and a One-A-Day. I eat a good diet, so the supplements are just for insurance. However, I believe in moderation. I don't think people should take massive doses of vitamins."

In poor countries, he said, a third of cancers result from chronic infections: hepatitis B and C and liver cancer, human papilloma virus and cervical cancer, *Helicobacter pylori* and stomach cancer, to name a few.

"The selection process in human evolutionary history favored reproduction and selected for mechanisms like white-cell defenses that protect humans during early life," he said. "But we paid a price for that protection in the form of increased cancer. Those white cells that guard against our demise from infection use nasty mutagens to incinerate their targets."

Microorganisms that cause chronic infections also kill cells and stimulate cell division, he said, and high rates of cell division are associated with increased mutagenesis and carcinogenesis.

Mutagens damage the DNA that makes up genes. Cancer results from multiple mutations, which means that most, if not all, mutagens are also potential carcinogens. Although cells have enzymes that continually cruise along the DNA and repair it by clipping genetic aberrations out of the lineup, as people age the mutation rate gradually exceeds the body's ability to remove these bad actors, giving cancer a chance to establish a foothold.

"What's amazing is how well defended we are against the damaging effects of chronic infections," Dr. Ames observed. "It takes 30 years or so to get those cancers."

Even if a cell's DNA is seriously damaged by a mutagen, however, cancer will not result unless that cell is actively dividing. Herein lies Dr. Ames's problem with animal tests using high doses of suspect chemicals. These "maximum tolerated doses" injure cells and tissues and set off reparative cell division. Dr. Ames maintains that in many cases it is the accelerated cell division, not the chemical per se, that is causing cancer. Stimulation of cell division may also account for the role of certain hormones in cancer, like the contribution of the body's estrogen to cancers of the breast and uterus.

But people are rarely exposed to anything like the large doses of chemicals used in animal tests, and at the more typical low doses that enter the body, there is no toxic damage to produce cell division and therefore they are not likely to increase cancer risk, he said.

"The dose makes the poison," Dr. Ames said, quoting a maxim of toxicology. "At some level, every chemical becomes toxic, but there are safe levels below that. You cannot extrapolate linearly from a high dose to a low dose and end up with a realistic risk estimate. A tenfold reduction in dose would produce much more than a tenfold reduction in cancer risk."

He pointed out that about 40 percent of chemicals that cause cancer in rodents at high doses are not mutagens, which suggests that the toxic dose, not the chemical per se, is often the problem.

In fact, Dr. Ames and others have shown that exposures to low levels of toxins induce natural defense mechanisms, including various enzymes, antioxidants and the melanin that forms in Sun-exposed skin. Then, when a larger dose of the insult is encountered, the defense systems are primed to act protectively, Dr. Ames said.

"Environmental pollutants are not an important cause of cancer," he said. "They account for a tiny percent of cancers in Americans, but might be a problem in people like farm workers who apply pesticides if they are heavily exposed. Hysteria about pollutants is costing us an enormous amount. We spent a huge amount of money getting traces of the metal degreaser trichloroethylene out of our water, which had a possible hazard 100 times below that of the average amount of chloroform in chlorinated tap water, although in fact neither may be a real hazard."

People fret about pesticide residues, but 99.99 percent of the pesticides Americans consume are natural constituents of plants, Dr. Ames said. He has calculated that the typical American eats about 1,500 milligrams a day of natural pesticides, which is 10,000 times the average daily consumption of 0.09 milligrams of synthetic pesticide residues.

Like the synthetic pesticides, about half the natural pesticides that have been tested are carcinogenic in rodents. In trying to reduce dependence on synthetic pesticides, some agricultural scientists are breeding plants with higher levels of natural pesticides, in effect trading one toxin for another, Dr. Ames said. For example, a new insect-resistant variety of celery was found to contain 6,200 parts per billion of carcinogenic chemicals called psoralens, whereas normal celery had only 800 parts per billion of these chemicals.

"On average, half of all chemicals tested, natural or synthetic, are carcinogenic in animals," Dr. Ames said, adding: "It is not true that our bodies are better able to defuse natural toxins than synthetic ones. Our defenses are general ones; they do not discriminate between natural and synthetic chemicals.

"I don't mean to suggest that there aren't real problems with some synthetic chemicals, but the environmentalists are wildly exaggerating the risks. If our resources are diverted from important things to unimportant things, this doesn't serve the public."

—JANE E. BRODY
July 1994

A LEVEL-HEADED LOOK AT RISK

Hardly a week passes without a new report that something in the foods we eat, the beverages we drink or the air we breathe contains a substance that causes cancer in laboratory animals. The anxiety and fear generated by such reports have prompted many consumers to wonder whether it is safe to eat fruits and vegetables that may have been treated with pesticides or grown in soil supplemented with fertilizer. Through the years, various poorly documented cancer scares have wreaked havoc in some industries—like the 1989 scare about Alar, a chemical that had been used to control the ripening of apples. The celebrity-promoted Alar scare resulted in millions of apples and countless gallons of apple juice being dumped by parents who feared for their children's safety. And apple growers nationwide suffered economically, some to the brink of bankruptcy.

But what do we actually know about the causes of cancer in people? We know that cigarette smoking, a practice chosen by 25 percent of American adults and an ever growing number of American teenagers, is by far the single leading cause of human cancers in the Western world. We know that diets too high in calories, too low in fiber and deficient in protective fruits and vegetables are responsible for at least a third of human cancers. We know that excessive consumption of alcohol can result in cancers of the upper digestive tract and liver. We even know that lack of exercise increases the risk of developing some cancers.

So why are so many concerned individuals and government agencies spending so much time, energy and money on unproven risks from substances to which few of us are more than minimally exposed? Because studies of risk perception have shown that people are far more fearful of potential dangers over which they have little or no control—things that are foisted upon them by industry or government or nature—than they are of self-chosen risks, even though the latter may be far more certain and potent hazards.

In 1987, Dr. Bruce Ames and his colleagues at the University of California at Berkeley sought to establish a more rational perspective for consumers and regulators about the possible causes of cancer in the environment. They devised a ranking system of risk based on the potency of each chemical in laboratory animals and how heavily people are likely to be exposed to it. The resulting index, which they called HERP (for human exposure/rodent potency), indicated, for example, that the cancer risk that might result from eating one peanut butter sandwich or one raw mushroom or drinking one cup of comfrey tea a day was a hundred times greater than that of daily exposure to the DDT residues in foods at that time.

Nature, as the HERP index demonstrated, is not benign. At least half of all natural chemicals, as well as half of synthetic ones that have been selected for testing, cause cancer in laboratory animals, who are exposed to doses far higher than would ever be encountered by the average consumer. Perhaps it is time, as Dr. Ames has repeatedly suggested over the years, to stop fretting over minor risks and pay a lot more attention to avoiding the major ones that people can readily control.

—J.E.B.

James D. Watson

Impresario of the Genome Looks Back With Candor

Vic DeLucia *for* The New York Times

Not every day do you see a renowned biologist dancing vigorously with an unclothed woman. But here indubitably is Francis Crick, his head thrown back in laughter, his arms around a partner who does not seem very intent on asking him about deoxyribonucleic acid.

The photograph comes from the manuscript of an author renowned for his unvarnished truths. He is James D. Watson, Dr. Crick's colleague in discovering the structure of DNA in 1953. The dancer, Dr. Watson explains as he leafs through the pages on the table of his office at the Cold Spring Harbor Laboratory, came from an agency as a long-ago birthday surprise. The manuscript is the second volume of Dr. Watson's autobiography.

The first volume, *The Double Helix*, published in 1968, caused a stir because it portrayed the personal lives of scientists as well as the century's most important discovery in biology. The sequel is about the cracking of the genetic code, which followed in a rush of discovery after the structure of DNA became known.

The publisher is not yet happy, however. "They want my mature reflections," Dr. Watson complained, "but I wrote it through the eyes of the callow youth I was then."

Honest Jim was the working title of his first book, signaling its unstinting candor about almost everyone involved. His friends may need to brace themselves again, or maybe Dr. Watson is holding back a little this time: "It's so difficult writing about living people," he said.

Candor may seem incompatible with tact, but in a contradiction that no one can quite explain, Dr. Watson is as effective as any diplomat. He has built up the Cold Spring Harbor Laboratory into one of the world's leading centers of molecular biology. He was the central figure in shaping the human genome project, the ambitious program of deciphering all three billion letters of human DNA by 2005.

Dr. Watson has, of course, been empowered by the greatness of the discovery that he, a 25-year-old biologist from Indiana University, made with Francis Crick at Cambridge University in England. Other biologists stand in awe of the fame attached to his name, some saying he will be remembered as long as Newton, and others comparing him with Darwin and Mendel.

"Every time you are in Watson's presence you realize are in contact with a man who has changed the course of human history and who will be remembered long after you have turned into dust," said Dr. Philip Sharp, a biologist at the Massachusetts Institute of Technology.

But Dr. Watson has used the power of his celebrity, critics complain, as a license to say whatever he thinks. Some biologists fear to disagree with him for fear of a thunderbolt impugning their intelligence or sanity. Dr. Watson does not hesitate to make invidious comparisons. He tells a visitor that a former associate is "two orders of magnitude, at least two orders of magnitude brighter" than a certain well-known biologist whose name has happened to come up in conversation.

A saving grace is that the power and the candor have generally been deployed toward ends that Dr. Watson's colleagues share. The thread that runs through his activities, as if to repay the source that ennobled him, is the promotion of molecular biology. But even molecular biology is not an end in itself. "It's not promoting biology for its own sake—it's because I am interested in the answers," he said.

Dr. Watson has let few things stand in the way of getting those answers. In his younger days, at the Harvard biological department in the 1950's and 60's, he offered no quarter to those who could not understand that the future of biology lay at the molecular level, not in the traditional studies of the whole animal.

The naturalist Edward O. Wilson was still seething over those attacks 30 years later, when he wrote in his recent autobiography: "At department meetings Watson radiated contempt in all directions. He shunned ordinary courtesy and polite conversation, evidently in the belief that they would only encourage the traditionalists to stay around. His bad manners were tolerated because of the greatness of the discovery he had made, and because of its gathering aftermath. . . . Watson, having risen to historic fame at an early age, became the Caligula of biology."

Dr. Watson's presence alone bespeaks an unordinary individual. He will scrunch his eyebrows over his ice-blue eyes, as if peering to see a visitor from a great distance. Then the eyebrows fly up, his whole face a mirror of surprise, as if he were himself dazzled by the clarity of a point he has just made. Today, he is not playing Caligula. Today is more Augustus, but gossipy. He talks of one scientist as a "rogue," who stole a great discovery. He tells of another, the administrator of a large institution, who brought a bishop to his inauguration, a sure tip-off to the embezzlement that followed.

It is time for a tour of the estate. Dr. Watson might have made Cold Spring Harbor a monument to himself, like Jonas Salk's soaring palace in La Jolla, California. Surprisingly, he has done the opposite. The scale is human, the focus on a movement, not a superhero.

Cold Spring Harbor used to be a little whaling village, set on the slopes around an inlet of Long Island Sound. That ambience has been preserved as much as possible. The houses have been elegantly restored, but are now home to laboratories. There are lawns overlooking the small harbor, with places to talk or think in beautiful surroundings. Dr. Watson and his wife, Elizabeth, an expert in historic restoration, pay attention to every detail. Today he fusses over the hanging of a new portrait of Milislav Demerec, a former director.

What is commemorated at Cold Spring Harbor is not a person but a moment in history, the birth of molecular biology. Photographs of the subject's founders hang on the walls. Most are not of individuals but of groups of people in discussion, as if to emphasize the communal nature of the biologists' search for knowledge, especially in its heroic early days. Dr. Watson walks through the meeting room where in 1953 he gave the first public description of the structure of DNA. There is a picture of him on the wall; he is in a group photo, his face so small he has to point himself out.

The laboratory, an institution with both bright and dark passages in its history, was slipping into decay when Dr. Watson became its director in 1968. He proved a skillful fund-raiser, adept at winning support from the local community as well as Federal sources. He hired many biologists of distinction, and exerted his influence beyond Cold Spring Harbor by convening conferences on emerging new topics and setting up training courses in critical new techniques. Rival institutions may do as well or better, but many biologists have spent some important moments of their careers at Cold Spring Harbor. If the world's molecular biologists acknowledge any particular home, it is the little hillside village that Dr. Watson has so carefully rebuilt.

Dr. Watson has also played a major role in shaping the Human Genome Project. "It was so obvious you had to do it," he said. But many of his colleagues at first needed much persuading of that thesis. The Department of Energy had started on the project in the late 1980's and biologists feared their research grants would be squeezed if their own financing agency, the National Institutes of Health, were to get involved. Dr. Watson brought his colleagues around, persuaded Congress to provide funds and became the first director of the National Institutes of Health's genome office in 1988. He enlisted high-quality scientists in the project, and specified that animal genomes were to be sequenced too so as to help interpret the human genome. "It was his stature and credibility that held things together," said Dr. Norton D. Zinder of Rockefeller University.

"I would only once have the opportunity," Dr. Watson wrote in his laboratory's annual report, "to let my scientific career encompass a path from the double helix to the three billion steps of the human genome."

Dr. Watson knows that the molecular biology he has nurtured is a force that could change the world. He chuckles at the embarrassment the physicist Stephen W. Hawking is said to have caused by talking at a White House lecture last month of the engineering of humans. Of course Dr. Hawking would want to improve the human design! Dr. Watson's eyebrows shoot up in amazement that anyone could fail to understand the yearnings of a highly intelligent physicist dependent on machines to talk and move.

To a degree that might surprise those who see him as a biology booster, Dr. Watson has always been sensitive to the misuse of genetics and to biologists' need to maintain the public's trust. When researchers first learned in 1974 how to move genes from one organism to another, he was among the signers of a famous letter that called for an unprecedented moratorium on the technique until its possible dangers were better understood. One of his first acts as director of the Human Genome Project was to announce that 3 percent (now 5 percent) of its funds would be set aside for studies of its ethical and legal consequences.

Dr. Watson is well aware of the calamities that ensued from the eugenics movement and how easily unfounded academic ideas can get translated into murderous policies. One of his predecessors as Cold Spring director, Dr. Charles B. Davenport, was a leader of American eugenics in the 1930's. Dr. Watson has long feared his critics would discover that fact and accuse him of being a closet eugenicist. It was Demerec, "a Serb," Dr. Watson notes as he appraises the new portrait, who finally closed down Cold Spring Harbor's eugenics program.

Dr. Watson shares no part of the eugenicists' belief in coercive policies against those deemed genetically undesirable. But he does believe strongly in the impending power of genetics to improve the human condition, if it is allowed to do so.

"If we could honestly promise young couples that we knew how to give them offspring with superior character," he wrote in the laboratory's annual report last year, "why should we assume they would decline? Common sense tells us that if scientists find ways to greatly improve human capabilities, there will no stopping the public from happily seizing them."

Over lunch, Dr. Watson is in a reflective mood. This is his 30th year at the laboratory. He is now president, having handed over day-to-day responsibilities to a director, Bruce Stillman. Soon he will be 70, but the event will take place during a month-long vacation in Australia, as if he wanted no one to make too much of it.

Perhaps because of working on his autobiography, the time with Dr. Crick is much on his mind. Some of his remarks seem at first a little barbed. He talks of their different taste in friends. He speculates what a difference might have been made in British science if Dr. Crick, like himself, had taken an active role in scientific politics. Dr. Crick, who now works at the Salk Institute in California, has remained active in research, particularly on subjects like the brain, and has for the most part avoided administrative work.

But this is just Dr. Watson's candor, his habit of amassing details about people's behavior and plumage like the bird-watcher he nearly became. It soon seems clear that he is talking about Dr. Crick with the mix of affection and rivalry reserved for an elder brother. Dr. Crick is 12 years older, perhaps the intellectual anchor in much of their discovery, although it was Dr. Watson who saw how the bases must fit together. "Francis was always so kind to me," Dr. Watson said. "He never tried to promote himself. He was just interested in solving problems."

Dr. Watson points across the harbor to where a line of yew trees is being hewn to half height. The yew hedge stands in front of his house, occluding its view of the harbor. Half the house is now in sight. Soon nothing will obscure Dr. Watson's full view of his demesne. Both he and Dr. Crick, in their different ways, have spent all their lives pursuing the unimpeded vista, the answers that will finally explain the nature of life as best as science can understand it.

—NICHOLAS WADE
April 1998

BIRTH OF THE DOUBLE HELIX

In 1951, at the age of 23, James Watson arrived at the Cavendish Laboratory in Cambridge, England, determined to study the structure of DNA. At the time few scientists understood the importance of the problem. One who did was Francis Crick, a physicist working for his Ph.D. on the structure of proteins.

The two scientists decided they could divine the structure of DNA by building Lego-like models. Their first attempt was a humiliating failure and the director of the Cavendish laboratory, Dr. Lawrence Bragg, ordered them to quit. Since Mr. Crick was supposed to working on proteins, and Dr. Watson had told his funding agency in Washington he was studying plant viruses, neither was in a position to object.

Dr. Watson adroitly seized the chance for them to get back into the game a few months later when Dr. Linus Pauling, Dr. Bragg's American rival, published a structure for DNA that was clearly in error. Arguing to Dr. Bragg that Dr. Pauling would soon recognize his mistake and find the right structure, Dr. Watson secured permission to resume model building.

Mr. Crick knew that DNA must be wound in a helix, probably a double one, but he and Dr. Watson had previously assumed the bases must point outward from the central DNA backbones. By the time of their second attempt to build a model, they had begun to think the bases might point inward from the DNA spirals, an approach that forced them to consider how the bases might fit together.

As it happened they now shared an office with Dr. Jerry Donahue, an American chemist who had worked with Dr. Pauling and was a leading expert on the bases, as the four different subunits of DNA are known. Dr. Donahue told them the structures shown in the textbooks were wrong, information that allowed them to make correctly shaped cardboard cutouts of the four bases—adenine, guanine, cytosine and thymine—known for short as A, G, C and T.

At about the same time they had belatedly become aware of Chargaff's rule, a curious experimental finding that in a wide range of species the amounts of A and T are the same, as are those of G and C.

Playing with the cutouts on his tiny desk, Dr. Watson tried to pair A with A and T with T, according to his conception of how the bases might fit together. Suddenly he noticed that an A-T pair he had idly made was identical in shape with a G-C pair lying nearby.

He at once realized that the observation would explain how matching pairs of bases pointing inward from the two spiral chains would form steps of equal width, provided that A always paired with T, and G with C.

The pairing also would explain Chargaff's rule and, more importantly, how one DNA chain could serve as the template for building another, an essential feature of a molecule that embodied hereditary information.

Mr. Crick was not more than halfway through the door, Dr. Watson later wrote in his book about the discovery, *The Double Helix,* "before I let loose that the answer to everything was in our hands."

Mr. Crick was doubtful as a matter of principle, but could maintain his skepticism for only a few minutes. They soon went across to the Eagle, the dingy pub at which they lunched, where Mr. Crick told everyone in earshot that they had discovered the secret of life.

They had indeed.

—N.W.

Rodolfo Llinás

Listening to the Conversation of Neurons

He might well be called the pied piper of the neurons. He lives with them day to day, he vacations with them in the summer, and in his hands, they dance.

Dr. Rodolfo Llinás, chief of physiology and neuroscience at the New York University School of Medicine, has spent almost 40 years studying brain cells. He is one of the leaders of modern neuroscience and a principal architect of a new view of neurons in human thinking and behavior. They are no longer viewed as little switches, or cables flashing messages across the brain.

Rather, they are quite complex, with something like minds of their own. And this has implications for everything from unconscious work inside the brain to conscious behavior.

"I have studied single cells all my life, essentially," Dr. Llinás said, "and they are not neutral. They have a point of view, a personality, so to speak."

He has coddled, probed and poked neurons from eels and pigeons, cats and rats, frogs, fish and alligators, not to mention humans. And each year in the summer, he takes his vacations at the Marine Biological Laboratory in Woods Hole, Massachusetts, with the biggest of his little companions: giant neurons of the squid, which are 100 times as large as those of humans.

He speaks of neurons as if they were crowds of friends, or attendants at a cocktail party. "The cells are not sitting about quietly, waiting for the sense data to come in and rouse them," Dr. Llinás said. "When I think about the way they act, the feeling I get is like that when I go into a store to buy something, and I encounter a group of clerks talking to each other. I try to get their attention, but they're talking about the weekend or something. That's what it is like. The neurons are working and the sense data is knocking on the door trying to get in."

He added, "No wonder humans are so weird and unpredictable! Our cells are disagreeing!"

Dr. Roger Traub, a neuroscientist at IBM's laboratories in Yorktown Heights, New York, where he is building computer models of human neuron networks, describes Dr. Llinás as "one of the great neuroscientists of the age."

"We think about the brain differently as a result of him," Dr. Traub said. "Some people do beautiful cell work in the laboratory. Others are great thinker-types. There are not many people who do both, and Llinás is one."

Dr. Walter Freeman, a neuroscientist at the University of California at Berkeley, said of Dr. Llinás, "He works on single cells, but he also has a broad grasp and a philosophical approach." Dr. Freeman said that Dr. Llinás's work added a new dimension to neuroscience. "It is lovely to see him reaching out beyond the single neuron," Dr. Freeman said. "It is refreshing, and perhaps it is a sort of reaction against the strong reductionism which still dominates the field."

While Dr. Llinás smiles and says he is "really just an old reductionist," meaning one who supports the notion that all thought and consciousness can be reduced to physical behavior of matter, it is some of his work that has brought a certain richness back into neuroscience's seemingly mechanical image of humans, colleagues say.

Since William James a century ago, the dominant mode of thought in the laboratory sciences of the mind has been what has been described as "reflexology." In this view, human brains are input-output machines that, while making very complex responses to the world, are still essentially responding to the demands of the world.

In a recent interview in his office on First Avenue in Manhattan, Dr. Llinás disagreed. "The brain is a prediction machine," he said. It is there to make elaborate mental maps of the world, he went on, a virtual environment inside the head that is reliable enough to enable humans to predict what lies ahead, both in space and time.

Most living things—plants, fungi and the like—have neither nerves nor brains. But those forms of life do not have to move. "When you move, you must have some predictive power," Dr. Llinás said. "You must have at least a simple image of what you are moving into."

To get a sense of what is in the world ahead, an array of sensors is needed. The most ancient type of sensor, so probably the most important type to have, is the detector that senses gravity or, put another way, tells what is up and what is down. There is even one organism that grows a ganglion when it is young and searching for a place to live, and then reabsorbs its mental tissue once it settles down to life as an unmoving adult sponge.

Dr. Llinás, 62, is 5 feet 5 inches tall and relatively thin, but he seems larger. His most striking feature, besides his easy smile, is his head of white hair that puts one in mind of Charles Lindbergh in a triumphant moment.

He was born in Colombia, into a well-off and well-educated family in Bogotá. His father was a professor of surgery, and his grandfather was a surgeon and neuropsychiatrist. Dr. Llinás has become a medical scholar of the classical mode, with both an M.D. and a Ph.D., and reads widely outside science. His favorite writers include Dostoyevsky, Shakespeare and Gabriel García Márquez, a fellow Colombian and a friend. But Dr. Llinás said his informal learning growing up in Bogotá may have been more important than his book learning.

"From very early on," he said, "I was thinking and talking about the mind, because that is what the subject was in our house." There was one important day in Dr. Llinás's education that he recalls even now. He would often see patients sitting in the waiting room of his grandfather's home office. One patient who had epilepsy collapsed in a seizure in the hallway, before little Rodolfo's eyes. "I was very impressed," Dr. Llinás said. "The poor man was salivating and twisting."

Agitated, he later asked his grandfather, "Why would a person do that?"

He does not want to do that, his grandfather replied.

"He doesn't want to?" Dr. Llinás recalled asking. "How can he do something he doesn't want to? Such an incredible event! I began to wonder, when I move, how do I know whether I want to or not?"

By the time he was a teenager, he had his own laboratory in the basement. He had not just the usual chemistry kit, but also had a little room where he did experiments on mice and rats with electrodes. "My mother kept after me, to make sure I was not unkind, but there it was, I was already started," he said.

From that time in 1951 in Colombia, Dr. Llinás has worked on essentially only one problem: how does the meat of the brain give rise to the spirits of the mind?

He received his bachelor's degree from Gimnasio Moderno and his doctor of medicine degree from Universidad Javeriana, both in Bogotá, before traveling to the United States to work as a post-doctoral fellow at the Massachusetts General Hospital and Harvard Medical School.

Among his most important discoveries in the next decades was that neurons have many unsuspected complexities. To begin, he found that neurons had different "channels" or receivers on different parts of their anatomy. Some are fast, some slow. Some are powerful, some weak. Some excite cells and some inhibit them.

What he studied was the cerebellum—the brain's tract of cells that are used to choreograph body movement. He made one of the most fundamental discoveries in the field when he found that there was one small knot of cells—called the inferior olive—that acts as a pacemaker for the cells in the cerebellum. This demonstrated that neurons were independent, operating at their own pace, rather than just reacting to stimuli from the world.

This required some new thinking about movement, Dr. Llinás said. "It was believed that the information coming into the brain and spinal cord organized the system so that the animal could walk. But experiments were done which showed that if you cut off all incoming information from the world, the animal still walked without hesitation," Dr. Llinás said.

It turns out that the cells are actually "walking" all the time, that is, they are active and producing tiny tremor-like movements of the muscles continuously. The cells have already decided to walk, and it is only second that they look to the outside for information on where to put the foot, or whether to keep it still.

"The following," Dr. Llinás said, "really explains in a very few words what the nervous system does: the spinal cord says walk; and the information coming in from the world says wait, the terrain is not even, there are stairs here, and you must lift your leg more."

The next step for Dr. Llinás was to imagine that such a rhythmic system might control not only the body's movements, but the body's thoughts.

In the thalamus he found another pacemaker, like the inferior olive, only this one may set the rhythm for conscious thinking. It is called the intralaminar nucleus.

Could thought, like movement, be active and automatic, rather than reactive? Does the brain generate "an active internal state—like walking for the cerebellum—which is then modulated by the senses?"

"This intrinsic state," Dr. Llinás says of the mental environment generated by active neurons, "has many names. If you are asleep, it is called dreaming. If you happen to be awake, and it's very strong, it is called daydreaming. If you are aware of what is happening outside at the same time, it is called thinking."

Essentially, the brain is a dreaming instrument. "At night, it dreams in the absence of sensory input, and it is free to be as crazy as it wants to be," he said. "Or in the day, it dreams in a more limited fashion because the senses serve to limit the types of images you can generate."

And then, says Dr. Llinás, as the body moves, the brain, or mind, experiences the world. Sensing cells pour data into the brain, which sifts the information and constructs continuously a kind of "living map" of the world, in Dr. Llinás's words, a "reality emulation."

The odd part of all this, he says, is self-delusion. Because the brain is actively making both facts and fantasies, Dr. Llinás said, sometimes "our mental maps get so elaborate we get lost in them, or we hide from even ourselves."

"Humans are probably the only self-delusional animal," he said.

This is a kind of neuroscience that even a novelist could love. And so Dr. Llinás speaks of his friend Mr. García Márquez as a model of cognition: he sits at his desk creating fictional landscapes not unlike those more or less factual ones others create for themselves every day.

—PHILIP J. HILTS
May 1997

THE BRAIN'S RHYTHM SECTION

The cells of the human brain, Dr. Rodolfo Llinás has long believed, must get the beat right and dance together, or trouble can result.

And at the end of 1999, he announced that he had found the part of the brain responsible for setting that rhythm, keeping many of the body's functions operating in harmony.

For decades, Dr. Llinás has laboriously tracked the activity of individual brain cells and then followed their behavior in groups. Such bundles of brain cells, he says, analyze information together and fire a signal for action at the same instant, rather like different instruments of an orchestra playing different notes but keeping the same beat.

In recent years he has focused in on the thalamus, a small area of the mid-brain that has traditionally been thought of as a relay station, passing messages from the body and lower brain centers to cortical regions responsible for higher functions like thinking.

What Dr. Llinás observed, however, is that the thalamus does more than just pass along information. Instead, once messages reach the thalamus, they are set to a rhythm—synchronized, like a piece of jazz, to match the "beat" in the "higher" centers where thinking, vision, hearing, body movement and other functions are directed and controlled.

Dr. Llinás suggested that these centers of thought and action in the higher brain need more than a flow of information from the outside world. They need that information provided in patterns compatible with their own action. If the information does not arrive in the proper form, brain cells in, say, the area that coordinates movement begin to flutter out of control, firing ineffectually rather than in concert with the surrounding cadres of cells.

A breakdown in any part of the thalamus, Dr. Llinás said, is likely to lead to failure of coordination somewhere in the brain, and thus, somewhere in the body, perhaps triggering an array of illnesses from Parkinson's disease to chronic debilitating pain to ringing in the ears.

There has long been evidence that surgical lesions to the thalamus or electrical stimulation of the organ by electrodes can help relieve some disorders, including depression, obsessive-compulsive disorder, chronic pain, Parkinson's disease, and a variety of others. But in the absence of any reason why this should work, such findings have appeared rather haphazard.

Dr. Llinás's theory is still uncertain, and arguments for and against it—as Dr. Llinás puts it, "both flowers and flak"—continue to emerge. But experiments to test the theory are being sketched out now.

Dr. Llinás, a new grandfather, said he was anticipating the results of the coming experiments with glee, almost as much as he was looking forward to his first visit with the baby.

—P.J.H.

Michael L. Dertouzos

Unlikely Warrior Leads the Charge for Simpler PC

Rick Friedman for The New York Times

Dr. Michael L. Dertouzos, head of the Massachusetts Institute of Technology's computer laboratory, is railing against personal computers. "You want to use them as boat anchors," he says. "People should really revolt. I think we should take up arms."

The question that springs to mind is "What do you mean, 'we'?" Dr. Dertouzos, 60, does not buy his computers on sale at CompUSA, or sit for 20 minutes on hold waiting for tech support to help him figure out how to hook up his new printer. He has been running the MIT laboratory, which now employs about 500 people, for more than 20 years. Many, if not most, of the people who created the Internet and today's computers and software came out of that laboratory, where Dr. Dertouzos and others now run the consortium that sets standards for the World Wide Web.

Nonetheless, a computer will not boot any faster, even for an MIT wizard, and in a book, *What Will Be* (HarperEdge), and in a recent interview in his MIT office, Dr. Dertouzos detailed what is wrong with the mix of machine and software that offers the average consumer checkbook programs, Internet access and headaches. In essence, he says, the personal computer is just too complicated. Among his specific complaints about the PC are these:

- It has too many features, too few of which are the ones the user wants at any given moment.

- There is too much to learn. Dr. Dertouzos says his software manuals for the programs on his desktop computer are equal in length to an entire multi-volume *Encyclopaedia Britannica*.

- The machines often take charge, overruling the desires of the users. Dr. Dertouzos offers the example of being forced to accept a time-consuming upgrade to airline reservation software over the Internet when what he really wanted to do was use the software to make an emergency airline reservation.

- Enormous programming effort and vastly increased sizes of, for instance, word processors are devoted to trivial improvements in appearance or presentation.

Calling today's machine "user friendly" because of its endless choice of fonts and screen patterns, Dr. Dertouzos writes, "is tantamount to dressing a chimpanzee in a green hospital gown and earnestly parading it as a surgeon."

It might seem to the average computer buyer that these are fighting words, that Dr. Dertouzos is some kind of turncoat, saying what other techies (the world of Dr. Dertouzos is divided into techies and humies, or humanists) fear to say. Not so.

"Today's machines are just too complex to be accessible," another MIT guru and sometime rival of Dr. Dertouzos, Dr. Nicholas Negroponte, complained, acting somewhat shocked, in his column in the July issue of *Wired* magazine, home of all things cool and computable. Dr. Negroponte, head of the university's media lab and author of *Being Digital,* wrote: "Is it time for a strike or a users' cartel? You bet it is."

And other experts are ready to chime in. "Personal computers have always been abysmal devices," said Dr. Paul Saffo, a former student of Dr. Dertouzos and head of the Institute for the Future, a nonprofit group in Menlo Park, California, that specializes in management consulting. Over time, he said, they have become much more powerful, but not any better.

Dr. Leonard Kleinrock, a member of the computer science department at the University of California at Los Angeles who is a longtime friend of Dr. Dertouzos, coined the term "feature shock" years ago to describe what happens when people get new computers all filled up with things that they never wanted in the first place.

Even Jonathan Roberts, director of product management for Windows at Microsoft, arguably the source of some of the complications Dr. Dertouzos and others deplore, is apologetic. The biggest complaint that Microsoft gets, Mr. Roberts said, is that "you have something that worked and then you install something else and what you had doesn't work."

The sour notes from Dr. Dertouzos are particularly surprising in that he is a thoroughgoing technological optimist. "Toys, toys, more toys!" is the motto of his laboratory, he says, and it is one he embraces at work and play. He drives a 12-cylinder BMW. He designs equipment for his 37-foot power boat.

Dr. Dertouzos traces his optimism partly to growing up in Athens during World War II, while his father, who was an admiral in the Greek Navy, was in Egypt. After the war, other problems seemed small. Besides, he said, "It's easy to be a pessimist." He came to the United States as a Fulbright scholar in 1954. "I ended up in Arkansas," he said. And, in a true display of optimism, he added, "It was a great way to enter America."

He received his Ph.D. in electrical engineering from MIT in 1964, after spending some time in research and development. He started teaching at the university immediately, and quickly founded a high-tech company making early computer terminals in 1968. He sold that company in 1974 when he became the head of the MIT laboratory. He now frequents the national and international halls of power in industry and government. He is currently co-chairman of the World Economic Forum, a business-financed research organization in Geneva.

Even when he is criticizing computers, Dr. Dertouzos likes to point to the new and wonderful things these machines do, like enable users to cruise the World Wide Web. And when he complains of the sad state of computers now, the critique is only a prelude to how good the future will be. In effect, he is saying that the emperor is naked and shivering—right now. But the court clothiers are aware of the danger of pneumonia, so next year's suit will really be warm and comfy.

Dr. Dertouzos looks forward, with varying degrees of enthusiasm, to a world of intelligent kitchens that plan menus, cook meals and tell people what is in the pantry, to smart cars, smart closets that pick out matching clothes, even to a smart sink that reminds people to use a dental pick because of their risk of periodontal disease. Fortunately, the sink would be difficult to design and very expensive. Underlying all these fantasies is the notion of computers' becoming both smarter and easier to use.

Some of his friends say that his political skills, which have made the MIT laboratory so successful, may have tempered some of the criticisms he levels in the book. And Dr. Dertouzos does not spend his time pointing the finger of blame.

While Dr. Kleinrock is quick to say that "anything Microsoft does makes it worse," Dr. Dertouzos concentrates on the fact that the industry is in its infancy. Dr. Dertouzos compares today's personal computer to the first airplanes, familiar from old films that show them laboring to escape the ground and then crashing in a heap. Computers crash, and are difficult to use, largely because despite rapid technological change, they are still very new.

Another problem he sees is that engineers and programmers love to add a new twist here and a new capability there. Mr. Roberts at Microsoft admits, "It's much more difficult to make the everyday easier than it is to add a new feature." But the new features sell, Dr. Dertouzos points out. The marketplace, in the shape of the consumer, buys bells and whistles. "We're the culprit in that," Dr. Dertouzos says.

Indeed, says Dr. Saffo, "we are unwitting co-conspirators" in a system in which poorly tested products are sent to market for the buyers to test. In a recent study done as part of the Computer Industry Policy Project at Stanford University, Abron Barr and William Miller came up with evidence to show what observers of the industry have said all along. To make money with software, a company must get its product on the market first, and fix it later. Early examples of software, like a new Word or Quicken, are called beta versions.

Dr. Saffo observed, "We ask users to pay for the privilege of being beta testers." And the users do pay. For the moment, the economics of computers are such, Dr. Saffo said, that "I would fire the CEO of any company that held off its product until it was ready."

The answer, Dr. Dertouzos said, is that consumers must demand simple, usable machines that work more like cars than science projects. Turn them on and drive. No one would buy a minivan that had a 600-page manual on how to drive it. Why should people do that with computers?

One way to achieve simplicity is to bury deeper what the computer does, so that, for instance, a user can ask it questions in common speech. At MIT, a system called Galaxy, being developed by a team under the lab's associate director, Dr. Victor Zue, can answer questions posed to it about weather, airline reservations and restaurants in particular locations. It has been in the works for 20 years and so far has cost about $30 million.

Another approach suggested by Dr. Dertouzos is to allow users to tailor their computers to themselves, allowing them, in effect, to become programmers. Of course, then the programming has to be simplified, or else things are made worse, not better.

Simplicity is a bit like marital fidelity, in that while everyone is in favor of it in principle, practice varies. Microsoft, for instance, says Mr. Roberts, is on the simplicity train. In the next version of Windows there will be a feature called "on now," which will make a computer come on right away, as they used to 15 years ago, instead of taking minutes, as they do now.

But another so-called improvement that Mr. Roberts mentioned seems to go directly against Dr. Dertouzos' ideas. Users will be able to tell "where a file was installed." Dr. Dertouzos' contention is that users do not even want to know words like "file" and "install."

"I'm really quite pessimistic," Dr. Kleinrock said, about improvements in personal computers. And Dr. Saffo said that in truth the problems were never solved. Technologies change so fast that they become obsolete before they mature. Computers are not likely to get simpler. Something unforeseen is likely to replace them.

For the moment, Dr. Saffo said, he is pleased to see Dr. Dertouzos criticizing his intellectual offspring, while he paints a rosy picture of the future. Dr. Dertouzos trained many of the people who have made the current computers what they are, Dr. Saffo said. "It's the sign of a good teacher to criticize the work of his students," he said. "He's playing exactly the role he should play."

—JAMES GORMAN
June 1997

SISYPHUS WOULD SYMPATHIZE

Some three years have passed since Michael Dertouzos decried the unwieldy state of computers in the accompanying profile. In the computer world, as any-

one with even a passing knowledge of the subject can attest, that is the equivalent of at least a lifetime (and certainly more than enough time for the average home computer to become seriously outdated).

Nevertheless, critics say, computer and software designers still seem intent on turning Thoreau's advice—"simplify, simplify!"—on its head: "complicate, complicate" seems to be the philosophy driving the industry.

The quest for simplified computers is "like chasing a kite string," says Dr. Paul Saffo, head of the Institute for the Future, a nonprofit research institute based in Silicon Valley. "Every time it feels like it is just about in grasp, the kite leaps ahead, the string leaps ahead, and we're right back where we started."

There are a number of reasons for this. Perhaps the most basic is that however much experts lament annoyances petty and serious that come with computer complexity, there is still not a loud enough outcry from the people who matter most: the consumers.

Like purchasers coming into a car showroom—and experts often like to contrast the development of the car with that of the PC—people who want to buy a computer are often not, in the end, attracted to the models that offer less. They generally want all they can afford, and more.

There have been some serious efforts to market Internet appliances, as some call them: essentially stripped-down commuters designed to surf the net. But inevitably, experts say, their owners want to upgrade them—and the trouble starts all over. "Stripped-down does not equate with ease of use," Dr. Saffo says. "It's a way of copping out of it."

Many still hold out hope that the computer experience will become a friendlier one. Some think it may result as the industry moves closer to adopting Linux, the operating system that some think could seriously undercut the position of the software giant Microsoft. Others think simplification will come only when PC's themselves fall by the wayside and are replaced by elaborate networks of user-friendly computers that will be integral to the running of the home.

But for now, the average computer experience mixes exasperation with its convenience. And one reason may be that the giants of the PC business seem to be more concerned with acting like businesses than with making their products easier to use. Much of Microsoft's attention has been taken up by its battle against the government's antitrust case, while other companies are forever in search of the megadeal, like the one that may join America Online with Time-Warner.

Not that these developments have been without their pleasures. Who—except, perhaps, Bill Gates—did not enjoy the spectacle of a top Microsoft executive testifying at the antitrust case and, as he attempted to make his point with a computer, running into a series of problems only too common to the average home user? At one point, the screen simply went blank.

But in general, the problems with computers are less an occasion for laughter than for tears. A year after this story ran, Dr. Dertouzos said he still had to contend with computer crashes at his lab. "I still kneel and cry when some situations arise," he said.

—Eric Nagourney

Anthony S. Fauci

Consummate Politician on the AIDS Front

Marty Katz for The New York Times

If everyone in the world were like Dr. Anthony S. Fauci, there would be no need for Prozac. By any sensible reckoning, the man should be wilting around the edges. He has been at the center of the AIDS tornado since the epidemic began, serving as the director of the National Institute for Allergy and Infectious Diseases and the Government's Office of AIDS Research, which together oversee much of the AIDS-related research carried out around the nation.

Apart from his official credentials, he has been the most visible and quotable spokesman on every medical, epidemiological and social aspect of the disease, the de facto AIDS czar until the appointment in 1993 of Kristine Gebbie as the czar de jure to the President.

The institute that Dr. Fauci runs at the National Institutes of Health controls not only about 40 percent of the Federal budget designated for AIDS, but also the money for studies of all the other infectious diseases afflicting the nation's citizenry, from tuberculosis to hantavirus infections. In addition, Dr. Fauci runs a laboratory on the Bethesda, Maryland, campus that does basic research on how the human immunodeficiency virus gradually disassembles the immune system, studies that lately have yielded insights into AIDS and buffed his luster as something more than a superbureaucrat.

Even among scientists, where work addicts are commonplace, Dr. Fauci, 53, is renowned as a truly hard worker. This is the bookend refrain for any discussion about him, an emphasis first and last on how astonishing it is that he manages to work 16 hours a day, day after day, year after year.

And he is working on a plague for which, scientific revelations about the virus notwithstanding, there have been depressingly few advances to help human patients. There is no vaccine in sight. There are no new drugs in the pipeline, and the antiviral drugs that do exist, like AZT, DDI and DDC, have been shown to be of questionable benefit in prolonging life.

The epidemic continues to spread relentlessly across the globe. It has killed 210,000 people in the United States alone and eviscerated entire subcultures, including the arts community, which gives depth and resonance to the rest of the population. It is a new disease that has turned the world old overnight.

Dr. Fauci has been bounced around by activists in the AIDS community, denounced one day, embraced as comrade and hero the next. Activists recently convinced Congress that the Office of AIDS Research should be removed from his jurisdiction and given a full-time director of its own.

Yet Dr. Fauci looks the same as he has looked for years, a compact, meticulous, supremely confident, unflappable bullet of a man, at once compassionate and hard-driving, gentle and ferocious. The man with the natty suits, hearty handshake and resilient Brooklyn accent.

Doesn't he ever doubt himself, or his ability to make a real dent in the disease? Doesn't he ever get tired or depressed or demoralized, or just overwhelmed by it all?

"If you're asking whether I have fears, pain and anxieties, yes, of course I do," he said. "That's a natural human thing. But do you mean, have I ever let them interfere with my responsibilities, or get in the way of my work? No."

Not once? Not one single lapse?

"Never," he said firmly. "I have the self-discipline not to let that happen." As though to make a point about his purposefulness, he jumps up from the couch where he has been sitting in his office and strides over to his desk. Two of the walls in the large room are covered practically to the ceiling with his plaques and awards and degrees real and honorary.

"To take on what I take on, you have to be very organized and energetic," he said. "I have made enormous sacrifices in my family and personal life. But I don't want to be praised or pitied for this. I do it because this is what I want to do."

He is religiously organized, his day blocked out in chunks for his administrative duties, his rounds to see AIDS patients in the research hospital at the institutes, his lab meetings, his discussions with scientists in his group. He runs seven miles every lunch hour regardless of the brutality of the weather. He sets aside every Saturday evening and all day Sunday for his family, a commitment that is essential if he is to see his three young daughters while they are awake.

And perhaps his is the attitude that must prevail in the plague years: not robotic, because robots break down, but calmly obsessive and matter-of-fact. His pragmatism may be his strongest suit, the impulse that ultimately overrides what some say is his sensitivity and a tendency to take things personally. When told that Dr. Harold Varmus, the new head of the NIH, had described him as "running his institute with an iron fist," Dr. Fauci made a point of asking every subordinate he encountered that evening whether the description was accurate. Most giggled nervously and said variations of, well, yes, but you're always fair.

Dr. Fauci is a man of many opinions, the most essential of which are scientific ones. And it is in the scientific arena that he is most optimistic, although cautiously so. "We've been on a roller-coaster ride for years, from great enthusiasm to despair," he said. "Recently, there was a sense that the sky was falling, when it became clear that giving people AZT early in the course of the disease doesn't give significant benefit.

"But despite the feeling of a roller-coaster ride," he added, "the science will march ahead to the time when we'll have an enduring solution or a therapeutic intervention. And it all will be based on a step-by-step understanding of the immune pathogenesis" that is the hallmark of AIDS.

Dr. Fauci has devoted most of his attention of late to dissecting those steps. In widely praised work published last year, Dr. Fauci and his colleagues reported that the human immunodeficiency virus never really lies latent in the body, but rather is sequestered in the lymph nodes, where it disturbs many threads of the body's densely interwoven immune system. It overexcites some immune signaling pathways, while eluding the detection of others. And though the main target of the virus appears to be the famed helper T-cells, or CD-4 cells, which it can infiltrate and kill, the virus also ends up stimulating the response of other immune cells so inappropriately that they eventually collapse from overwork or confusion.

He proposes that one possible approach to treating infection with HIV is not to focus on clearing every last viral particle from the body, but rather to figure out exactly how the immune system comes unstrung, and then replace or compensate for those signaling pathways thrown out of whack. For example, he and his colleagues have learned that AIDS patients suffer from a defect in the arousal of an immune signaling molecule called interleukin-2. In trials now under way, they are giving patients IL-2 in carefully calibrated doses designed to recapitulate the activity of a normal immune system. They have found that the patients' T-cell counts improve with the treatment, but they now must determine whether that rise in cell number translates into an improvement in symptoms of the disease; for example, whether the interleukin treatment causes Kaposi's sarcoma lesions to shrink.

Dr. Fauci and his colleagues are also studying the immune systems of 10 people who are members of a small and fortunate minority: they have been infected with HIV for 12 years or longer, yet they have not progressed to showing any symptoms of AIDS. "We think the real clues to the disease are with these people," said Dr. Fauci. "What is it about them that's different? The architecture of their lymphoid tissue appears to be preserved compared to those who progress, and we're trying to see what is responsible for that preservation."

Dr. Fauci's current fixation on his science is what keeps him from getting excessively rankled by recent struggles over who will have the greatest hand in shaping the course of AIDS research. The legislation that reorganized the Office of AIDS Research last spring gave the office more strength than it had had under Dr. Fauci, with control over the entire $1.3 billion budget for AIDS research, keeping track of all AIDS-related programs and clinical trials that currently are run by 21 different institutes at the NIH.

Dr. Fauci and many other scientists had vigorously opposed the reorganization of the Office of AIDS Research, arguing, among other things, that it would emphasize a trend towards excessively targeted science at the expense of untethered basic research.

Dr. Fauci soft-pedals the degree to which his pride has been wounded, but friends and colleagues said they sensed his frustration. "It might be a relief for him to get some of the pressure off his back," said Dr. Donald Frederickson, a former director of the NIH and one of Dr. Fauci's old friends. "It's enough to do to run his own institute, and his lab is now going full steam. But nobody likes to be downgraded from the position of national importance that he's had."

Dr. Fauci, say friends and co-workers, has always been like a child's punching bag, which can be pushed over but will always pop back up for more. For the first eight years of his directorship at the allergy and infectious diseases institute, he worked under Republican Administrations reluctant to focus their attention on AIDS. He has been the repeated target of activists furious at what they perceived to be bureaucratic indifference and timidity—and he has also been their greatest friend.

"I call him murderer or hero, depending on the week," said Larry Kramer, a playwright, journalist and the most resilient bullhorn of all activists. "It's been such a complicated relationship." That complexity was reflected in Mr. Kramer's play, "The Destiny of Me," in which Dr. Fauci is portrayed, with scant attempt at disguise, as Dr. Anthony della Vida, or Dr. Life, a chipper physician-researcher who accepts descriptions of himself as "a son of a bitch of Hitler" and a "scientific fraud" from his churlish AIDS patient—modeled after Mr. Kramer—yet who tries with equal fortitude to keep that patient alive. Dr. Fauci attended the play's opening.

"I've worked with Dr. Fauci for years, and during that time we've had a combination adversarial and collaborative relationship," said Derek Hodel of the AIDS Action Council, a Washington-based lobbying group for AIDS organizations. He and other activists give Dr. Fauci tremendous credit for lending his support several years ago to the idea of a parallel track, of allowing experimental drugs to be made available to AIDS patients even as the compounds are tested in clinical trials. They also appreciate that he has spoken out against mandatory HIV testing of health care workers, and has opposed barring immigrants who carry the virus. They said he has been willing to listen to members of the AIDS community and to persuade grudging researchers and pharmaceutical companies that they should do the same.

Yet for every cheer there is a hiss. Some activists recall bitterly that during the Reagan and Bush Administrations Dr. Fauci defended research budgets that they thought were dreadfully inadequate. Some say he is a great scientist but a poor administrator; others insist that he is too smooth a bureaucrat to make any scientific headway. Scientists also have criticized Dr. Fauci as capitulating to the demands of activists too readily.

Through it all, Dr. Fauci accepts the criticisms, and he accepts that someone must absorb the anger and terror that AIDS has spawned, so why not somebody of strong vertebrae who was raised on the streets of Bensonhurst? "I was on a C-Span program a couple of months ago with Tony, and I attacked him for the entire hour," said Mr. Kramer. "He called me up afterwards and said he thought the program went very well. I said, 'How can you say that? I did nothing but yell at you.' He said, 'You don't realize that you can say things I can't. It doesn't mean I don't agree with you.'"

Dr. Fauci claims he does not take the intermittent blasts personally. "That's the activist mode," he said. "When there's a disagreement their tendency is to trash somebody. But I know that when Larry Kramer says the reason we're all in so much trouble is because of Tony Fauci, he's too smart to believe that. I don't want them to change or compromise that mode, as long as they don't ask me to change my opinions."

"I'm totally obsessed by this problem," said Dr. Fauci. "I'm going to finish this job." That is why he turned down an offer to take over the entire National Institutes of Health several years ago, he said, and that is why he does not mind ceding control over the Office of AIDS Research now.

"I think Tony believes he will find a cure for AIDS," said Mr. Hodel. "I hope he's right. And I hope a lot of other people are given the opportunity to believe that as well."

—NATALIE ANGIER
February 1994

HOPE RESHAPES AN EPIDEMIC

With a disease as complex and baffling as AIDS, few questions can be expected to have simple answers, not even the most important question: is there a cure yet? The short answer, of course, is no. Yet it is more complicated than that.

When Dr. Anthony Fauci was interviewed for this profile in 1994, a diagnosis of infection with HIV—the virus that causes AIDS—was tantamount to a death sentence. Although research into both treatment and vaccine was taking place on many fronts, with a vast commitment of private and government money supporting the effort, there seemed little reason to be hopeful for the near future.

Today, contracting HIV is still grim news. But now patients can expect to face years—perhaps many years—of expensive, taxing and sometimes arduous drug treatments. With the new medications for AIDS, life expectancy has been greatly extended, and symptoms significantly reduced, for people infected with HIV. Dr. Jerome Groopman, a leading AIDS experts at Harvard, recalls that when the disease was first identified in 1981, a patient could expect to live for 7 to 14 months after being diagnosed with full-blown AIDS. "The average survival now is unknown," he says.

Thanks in large part to new drugs—usually administered as a "cocktail" of as many as 30 pills over the day of drugs like AZT and protease inhibitors—the overall death rate from AIDS in the United States has dropped by about 90 percent in recent years. In the course of a single year, from 1996 to 1997, the disease went from being the eighth-biggest killer to the fourteenth.

But whatever progress has been made in the United States, the disease remains a scourge in the rest of the world, especially in developing countries. In Africa, for example, more than 2 million people died of AIDS in 1998. Around the world, more than 11 million children have been orphaned by AIDS—the vast majority in sub-Saharan countries.

And even in America, perhaps as people become more complacent in the face of medical advances, the HIV infection rate has remained fairly constant at about 40,000 cases a year, and there are signs that many people are failing to protect themselves from infection. In addition, the face of AIDS is changing. In 1998, for the first time since the virus was discovered, AIDS was diagnosed in more black and Hispanic gay men than white gay men.

Throughout the epidemic, the search for a cure has been far more complex than finding remedies for other diseases, and not just because the virus has proven so wily.

AIDS researchers have been plunged into a thicket of side issues, including pressure to make experimental drugs, however unproven, available to patients more quickly. Also draining have been battles over experiments involving infected people in the developing world, where researchers have been trying to determine just how little medication is needed to treat symptoms and to prevent the spread of the virus from mother to newborn. The goal is to make treatment as widespread as possible in cash-strapped countries, but because the experiments have entailed giving patients less than the standard doses of drugs used in developed countries, critics have raised ethical concerns.

As for a vaccine: three-year tests began on volunteers in 1998, but few experts seem to expect success in the short term. Most experts believe the key to controlling AIDS is going to be treatment. Dr. Fauci, who is still head of the allergy and infectious diseases institute, remains ever the optimist. In 1998, when figures showed that the number of Americans who died of AIDS had dropped to 16,865 the year before—almost half the number in 1996, and far below the peak of 43,000 in 1995—he said recent advances in treatment were encouraging.

"We now have the ammunition and a major positive impact on HIV-infected individuals when it comes to longevity and quality of life," Dr. Fauci said.

—Eric Nagourney

Lene Vestergaard Hau

She Puts the Brakes on Light

Rick Friedman *for* The New York Times

While riding her bike around Cape Cod, she could pass as a college freshman. But Dr. Lene Vestergaard Hau is no freshman. A teacher at Harvard University and an accomplished experimenter, she has gained a formidable reputation in a branch of science traditionally dominated by men: physics.

Most of all, she is known as the scientist who slowed light down to a walk.

Dr. Lene (pronounced LEE-nuh) Hau, at 39, is rapidly reaching the top of her profession, her colleagues agree. On February 18, 1999, the presti-

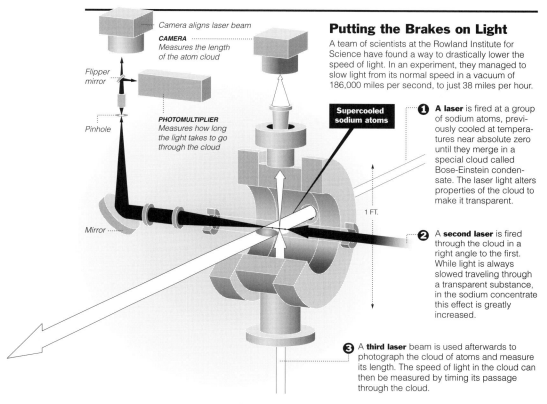

Putting the Brakes on Light

A team of scientists at the Rowland Institute for Science have found a way to drastically lower the speed of light. In an experiment, they managed to slow light from its normal speed in a vacuum of 186,000 miles per second, to just 38 miles per hour.

Camera aligns laser beam

CAMERA *Measures the length of the atom cloud*

Flipper mirror

Pinhole

PHOTOMULTIPLIER *Measures how long the light takes to go through the cloud*

Mirror

Supercooled sodium atoms

1 FT.

❶ **A laser** is fired at a group of sodium atoms, previously cooled at temperatures near absolute zero until they merge in a special cloud called Bose-Einstein condensate. The laser light alters properties of the cloud to make it transparent.

❷ A **second laser** is fired through the cloud in a right angle to the first. While light is always slowed traveling through a transparent substance, in the sodium concentrate this effect is greatly increased.

❸ A **third laser** beam is used afterwards to photograph the cloud of atoms and measure its length. The speed of light in the cloud can then be measured by timing its passage through the cloud.

Source: Lene Vestergaard Hau, Rowland Institute for Science

The New York Times

gious scientific journal *Nature* selected for its cover article a paper of which she was the leading author, and it caused a mild sensation.

In their paper, Dr. Hau, with Dr. Steve E. Harris of Stanford University and two of Dr. Hau's Harvard students, reported the results of their experiment in which a beam of laser light was slowed to the astonishingly low speed of 38 miles an hour. (By comparison, light in a vacuum travels about 186,000 miles per second.)

Dr. Hau's laboratory at the Rowland Institute for Science in Cambridge (where she conducts research with the help of her graduate and postdoctoral students from Harvard) is one of a handful of organizations studying the interactions of lasers with a very peculiar kind of matter called a Bose-Einstein condensate.

It was by shining precisely tuned lasers on such a condensate, or cloud, of ultra-cold sodium atoms that Dr. Hau and her team reduced the speed of a light beam to a pace slower than her bicycle.

The achievement, noted by many newspapers, magazines and broadcasters, was a tour de force of pure physics, and it may also herald great practical applications. Among them could be the development of optical switches that could enormously improve the performance of computers.

Dr. Hau presides over a half million dollars' worth of apparatus at the Rowland Institute, a research organization founded by the late Edwin H. Land, the inventor of Polaroid instant photography, to support innovative research. Unmarried, Dr. Hau lives near the laboratory and spends most of her time either at the laboratory or pondering the problems her next experiment will present.

"I find that the shower is a very good place to think things out," she said, "but sometimes I forget what I'm doing. Once I even stepped into the shower with my clothes on."

Although more women have entered the sciences in recent years, their participation in physics remains very limited. Of several hundred scientists attending a typical physics meeting, only a half dozen or so are likely to be women. Dr. Hau is often asked how she came to be one of them.

"For one thing," she said, "I was lucky to be a Dane. Denmark has a long scientific tradition that included the great Niels Bohr, one of the founders of quantum theory. In Denmark, physics is widely respected by laymen as well as scientists, and laymen contribute to physics. For instance, research in quantum mechanics has been supported in Denmark by the makers of Carlsberg beer since the 1920's. I myself was supported as a graduate student for one year by a Carlsberg scholarship."

Dr. Hau was born in Vejle, a small town not far from the University of Aarhus, on the east coast of Jutland. "Neither of my parents had any background in science," she said. "My father was in the heating business and my mother worked in a store. But both of them believed in giving me the same advantages as my brother, which was very important to my education."

A prodigy who skipped 10th grade, she was immediately accepted at a "gymnasium," a European upper school roughly equivalent to the first two years of college in the United States. From childhood Dr. Hau was fascinated by mathematics, especially geometry. "All my life I have needed to visualize things, even abstractions," she said. "Without a visualization in my head I'm lost, and geometry is very visual."

This approach to physics works, she said, even for visualizing mathematical abstractions like Hilbert spaces (named for David Hilbert, a German mathematician), which combine a variety of physical states (or "wave functions") into a single mathematical entity.

"When I first entered Aarhus University," Dr. Hau recalled, "I was bored by physics. They just taught us thermodynamics and classical mechanics, and that bored me. But I loved mathematics. I would rather do mathematics than go to the movies in those days. But after a while I discovered quantum mechanics, and that got me interested in physics again, and I've been hooked ever since."

Quantum mechanics is a system of probabilistic rules governing the discontinuous jumps in behavior of very small particles of matter and energy. Transistors and many other electronic devices work according to these rules.

Dr. Hau was awarded her Ph.D. from the University of Aarhus in 1991 after completing a dissertation on the "channeling" of electrons along strings of atoms in a silicon crystal, as if the atomic strings were miniature waveguides, like the optical fibers used to guide light.

Along the way she acquired a working knowledge of English, German and French. French proved to be vital to her work during seven months of research she conducted at CERN, the European Laboratory for Particle Physics, near Geneva. "The scientific papers were written in English, but the meetings announcing the availability of various particle beams were all in French, and without French I would have been lost," Dr. Hau said. "I loved it at CERN. The people were really passionate about their work. Even at 3 in the morning you would always see people in the labs."

In 1988, while completing her Ph.D. work, Dr. Hau made her first trip to the United States, looking for a post-doctoral job. By then she had received a one-year stipend from Carlsberg so she could accept a post-doctoral university appointment without pay. Among the scientists she met was Dr. Jene A. Golovchenko, a physicist at Harvard, who also worked at the Rowland Institute.

"I told Jene what I had been doing and also that I wanted a complete change in direction. I wanted to work on cooling atoms," she said. "He told me he didn't know anything about cooling atoms but said we could work together on it, so I was given a post-doctoral appointment. Later, the Rowland Institute gave me a staff job and my own laboratory."

Although she is a permanent United States resident, Dr. Hau has not applied for citizenship. "I would like to be a citizen so I could vote," she said, "but I would have to give up Danish citizenship, and that would make me feel disloyal. Last week the Danish comedian and pianist Victor Borge performed in Boston for two straight hours—and it was hilarious—even though he is 90 years old. Now there's a Dane for you, one to make you feel proud."

Although much of Dr. Hau's work is done on a blackboard, she is also a "dirty hands" physicist who builds and adjusts much of her own complex apparatus. In 1994, she and Dr. Golovchenko designed and built an ingenious atomic beam source called the "candlestick," which is now a part of Dr. Lau's light-slowing apparatus. The device incorporates a "wick" made of gold-plated stainless-steel cloth, which soaks up hot molten sodium metal and wicks the sodium up to a heater that vaporizes it. A jet of hot sodium atoms then shoots out of a pinhole into the cooling apparatus that chills them to a tiny fraction of a degree above absolute zero.

An atom at room temperature moves at high speed, but when it is bombarded from three directions by laser beams it loses energy and slows down: it cools off. In a complicated series of stages, Dr. Hau's apparatus uses lasers to cool the sodium atoms partway, and then evaporates the fastest (and therefore hottest) of them, saving the coolest in the trap. At the end of this cooling operation (which takes 38 seconds), the cloud of atoms in the trap has been reduced to only 50 one-billionths of a degree above absolute zero, a temperature far colder than any in nature, even in the depths of space.

Because of the Uncertainty Principle—one of the fundamental rules of quantum mechanics—the more precisely the momentum (or velocity) of a particle is known, the less precisely it is possible to measure its position, and vice versa. At exactly zero degrees (which in practice could never be reached), a particle would have zero momentum, a precise value. This would mean that the particle's position would be highly uncertain; it might be found anywhere within a large volume of possible places.

The volume occupied by each ultra-cold sodium atom in Dr. Hau's trap expands enormously, so much so that atoms in the trap are forced to overlap and merge into what physicists call a Bose-Einstein condensate (named for the theorists Satyendra Nath Bose and Albert Einstein), in which the atoms' quantum wave functions are combined.

Once the condensate is created, a "coupling" laser tuned to resonate with the trapped mass of atoms is beamed into the trap chamber so that the atoms and photons of light become "entangled" with each other, behaving as if they were a single entity. A pulsed laser probe is then shot into the "laser dressed" condensate from a different direction, and some of its light passes through, but at a speed 20 million times as slow as the speed of light in a vacuum.

A table covered with a labyrinth of little mirrors guides a network of beams requiring perfect adjustment, and Dr. Hau and her team work long and tedious hours, sometimes around the clock. Their eyes become strained from exposure to the glaring yellow laser spots reflected from the apparatus.

"Of course, we have to eat while we work, and we like pepperoni pizza, but we have to be careful to keep the optical system clean," she said. "Each cooling cycle for the sodium atoms takes 38 seconds, and that gives us just enough time to take a bite of pizza, change hands and flip up one of the mirrors in the optical system to prepare for the next cooling cycle. It takes some practice."

Dr. Hau is considering an experiment to measure localized wave functions in the cooled atom trap, allowing her to explore the structure of the condensate.

"Some people think it might be dangerous, that the trap might explode. But the most worthwhile physics is often a bit dangerous," she said.

"When you're looking at the computer monitor and you suddenly see what you had hoped you would find," she added, "it's the greatest feeling in the world."

—MALCOLM W. BROWNE
March 1999

SUCCESS'S LUXURIOUS HEADACHE

As Dr. Lene Hau slows pulses of light to an ever more leisurely crawl, her own life has sped up, leaving little time for anything outside her laboratory.

After rebuilding the laser apparatus with which she succeeded in slowing light to a speed of 38 miles an hour, Dr. Hau devised ways for reducing the speed of light still farther, to a mere 50 centimeters a second—about 1 mile an hour.

"We did it by incorporating a second laser in the system, one that has a slightly different wavelength from the first beam," she said. "And we're not finished. Our goal is to reduce the speed of light to one centimeter per second. A lot of insects can crawl that fast."

Dr. Hau became something of a celebrity in the spring of 1999 after she and a colleague, Dr. Steve E. Harris of Stanford University, published a paper in the journal *Nature* reporting the astonishingly low speed to which they had slowed pulses of light. (The normal speed of light is about 186,000 miles per second.)

"After all the publicity," she said with a chuckle, "I ended up with a very luxurious headache—a whole bunch of good job offers from great universities. It was really hard to decide. But in the end I picked Harvard. It's been a very hectic time with big decisions to make, but very exciting."

Besides a prestigious faculty position at Harvard, in which she will direct research by graduate and post-doctoral students, Dr. Hau was presented with her own laboratory.

"It's been a new and fascinating experience for me to work with architects in the design of shelves, the laboratory layout, even the color of curtains," she said.

Her voice warms as she talks about her research. "What I think will happen when we get the light moving slowly enough, is that atoms in the Bose-Einstein condensate will start surfing on the front of each tiny light pulse, moving very slowly, giving us plenty of time to manipulate things."

A Bose-Einstein condensate is a bizarre clump of intertwined laser light and atoms—sodium atoms, in Dr. Hau's experiments. She and her assistants are developing ways to probe condensates and explain their behavior.

"We have already created a fantastic new optical medium with great non-linear properties. That means we could build optical switches for computers and other things, so sensitive they could open or close under the control of just a single photon of light," Dr. Hau said.

Her meteoric success comes at some cost. Dr. Hau is often in her laboratory until late at night. "I'm not getting much sleep," she said. "I had hoped to take just a little time off during the summer, but there was never a chance."

"But I do get to do some things in my personal life. I've bought a condo in a big old Victorian house in Cambridge, and I'm getting ready to move from Boston. Cambridge is a lovely place, with lots of intellectual stimulation.

"I'm so happy!"

—M.W.B.

Carlos Cordon-Cardo

Cancer Trailblazer Follows the Genetic Fingerprints

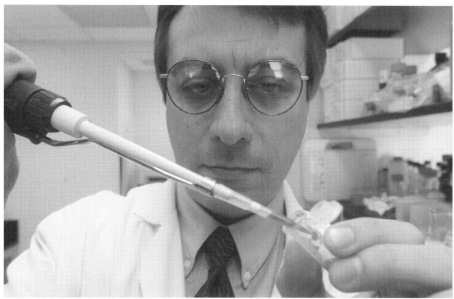

Dan Hogan Charles/The New York Times

Cancer was once thought to be a single disease that attacked different tissues. Then it was recognized to be a hundred different diseases as determined by the particular tissue involved and the kind of cells within that tissue that were growing out of control.

Now, the sophisticated techniques of a burgeoning field called molecular pathology are suggesting that cancer comes in countless variations, each with a genetically determined molecular fingerprint that indicates how deadly or indolent it is likely to be and how aggressively it should be treated.

"There are probably a million cancers, maybe as many as there are patients with cancer," said Dr. Carlos Cordon-Cardo, director of the division of molecular pathology at Memorial Sloan-Kettering Cancer Center in New York. "While there will be some garden varieties, we need to look at cancer as an individualized disease and develop individual treatment protocols for patients with different molecular profiles."

Dr. Cordon-Cardo, whose interest in pathology dates from his boyhood in Spain, is a leader among researchers hoping to use molecular pathology to transform the treatment of many cancers, increasing patients' chances of cure in some cases and avoiding unnecessary treatment in others. It is a hot field that is already beginning to alter the treatment of some common cancers, including breast cancer.

"Molecular pathology will revolutionize our approach to cancer diagnosis and treatment," said Dr. Richard Klausner, director of the National Cancer Institute. "We've been looking at cancer cells under the microscope for 100 years. We now know it's not the appearance of these cells but their molecular instructions that determine how a tumor will behave." He said the task of the next three to five years would be to determine which of the 50,000 genes that define a cancer cell are relevant to prognosis and treatment.

The new attention is welcome to Dr. Cordon-Cardo, who enjoys teaching lay audiences about the molecular determinants of the disease that has been the focus of his life's work.

"Two tumors may look identical under the microscope, yet one may go on for years without spreading and the other might kill the patient in a few years," he said. By linking different molecular markers with degrees of aggressiveness, it should be possible to tailor treatment to maximize the chance for a cure in patients with more virulent cancers and minimize the trauma of treatment for those with less aggressive cancers.

Already, researchers have discovered that cancers have types just the way blood does, only more of them. Analysis of the genetic characteristics of tumors has already begun to transform the diagnosis, prognosis and treatment of some cancers and is expected to radically alter treatment decisions for many other cancers in the years to come.

For example, researchers have discovered that 30 percent of breast cancers produce unusually high amounts of a marker protein called HER-2/neu, which renders them particularly ferocious. Researchers designed a drug called Herceptin that interferes with the marker protein and helps to control advanced breast cancer. The drug is now being tested against less advanced breast cancers as well as the 20 percent of ovarian cancers and some prostate cancers that produce too much of the protein.

Another success story involves the estrogen-receptor marker in breast cancer and the consequent use of the antiestrogen tamoxifen to reduce both the risk of recurrence and the development of a second cancer in the opposite breast. Molecular markers are also being used successfully to make treatment decisions in childhood leukemia and colon cancer.

The findings that are emerging almost weekly from molecular pathology laboratories are leading many to dismiss the notion that a single "magic bullet" would one day be found to cure all cancers.

While that may sound like bad news, Dr. Cordon-Cardo and others say the new findings are among the most hopeful developments in cancer research. They hold the promise of leading to new and more specific tools for early detection and for choosing treatment regimens that are just potent enough to eradicate a cancer but not any more powerful than necessary.

Through an understanding of a tumor's molecular fingerprints, Dr. Cordon-Cardo explained, it should be possible to avoid using the therapeutic equivalent of a sledgehammer to kill an ant when the cancer being treated is not particularly virulent. In other cases, when a tumor has fingerprints that indicate it is highly aggressive, oncologists may be able to tell in advance that they must throw the therapeutic book at it.

For example, prostate cancers vary greatly in their virulence, but doctors are unable to say for sure who will do well with minimal treatment and who will die without radical surgery or other aggressive therapies.

"Molecular signatures will allow us to identify tumors that are insensitive to a particular form of treatment from the get-go," Dr. Cordon-Cardo said. "Some of the genes we're studying will help us know in advance of treatment how certain drugs will impact tumors." Just as antibiotics are chosen based on the sensitivity of bacteria, chemotherapy or other therapies could be individualized depending upon the sensitivities of the cancer cells as determined by their molecular markers.

"For example, if the enzyme thymidylate synthase is increased in production in colon cancer cells, the standard drug 5-FU will not help because the cells are resistant to it," Dr. Cordon-Cardo said, citing studies by Dr. Joseph Bertino at Memorial. "There are quite a number of patients now being treated with 5-FU who may not be helped by it." Studies are now under way to determine if every patient with colon cancer should be tested for this marker before starting on 5-FU.

In a study published in January in *The Journal of Urology*, Dr. Harry W. Herr and other colleagues of Dr. Cordon-Cardo reported that chemotherapy and not surgical removal of the bladder might be all that was needed to cure patients with bladder cancer whose cancers were confined to the bladder and who lacked detectable levels of an important marker protein called p53. All 16 patients in the study who met these criteria were still alive a decade after treatment that preserved their bladders. On the other hand, those with tumors that expressed p53 responded best if their bladders were removed.

"We're at the point where organ preservation in cancer treatment is an important quality-of-life issue, and these kinds of tumor markers can help us determine when we can avoid removing an organ," Dr. Cordon-Cardo said.

P53 is a protein produced by a tumor suppressor gene that can have hundreds of different mutations, some of which are worse than others. "By testing for p53 mutations, we can tell which patients with such cancers as bladder, colon, breast and lung will do poorly and who won't respond to certain treatments," Dr. Cordon-Cardo said.

Also highly promising are studies of a protein called p27, which seems to play an important role in controlling the growth of many cancers. Studies in Boston, Seattle, Toronto, England and Japan have linked abnormally low levels of the p27 protein to aggressive cancers of the breast, colon, lung, esophagus and tongue.

Studies by Dr. Cordon-Cardo and colleagues indicate that p27 tests may help clinicians determine which prostate cancers are likely to be deadly and thus warrant aggressive treatment and which ones can be controlled with less treatment or perhaps none at all. He explained that p27 acts like a brake that keeps cell growth in check. When levels of this protein are low or absent, cells are likely to proliferate rapidly. Rapid cell growth, in turn, increases the chances of genetic mutations that may allow a cancer to develop and grow unchecked.

Another highly promising approach to have grown out of immunopathology and molecular pathology is the development and use of tumor vaccines to attack the abnormal proteins in cancer cells. Several such vaccines are beginning to be tested in patients, Dr. Cordon-Cardo said. For example, he and others at Memorial Sloan-Kettering have devised vaccines to boost the body's immune response to melanomas and neuroblastomas. The researchers inject molecules called gangliosides that spark the immune system into attacking similar molecules they have identified on the membranes of the tumor cells.

The achievements Dr. Cordon-Cardo and his colleagues are now forging were unimaginable when as a 10-year-old boy in Barcelona he asked for a microscope for his birthday.

"My grandmother, who lived with us, set up a corner of the house for me as a laboratory," he recalled. "I knew by age 17 that I wanted to study the basis of cancer, that I wanted to be a pathologist rather than a clinician. My mentor in medical school, Professor Lorenzo Galindo, spent 25 years in the United States and Puerto Rico and introduced surgical pathology to Spain. He told me I had to learn English and go to the United States to study."

A special fellowship from the Spanish Government enabled Dr. Cordon-Cardo to get a Ph.D. in cell biology and genetics at Cornell University Medical School. Then Memorial Sloan-Kettering asked him to stay on, giving him a small laboratory to do immunopathology, developing and characterizing antibodies that could help determine what biological changes allow a normal cell to become an immortal cancer cell.

"Within 15 years, we grew from a very small laboratory to a couple of laboratories to a division which now has 45 people," Dr. Cordon-Cardo said.

Meanwhile, a technological revolution has begun offering molecular pathologists the ability to automatically analyze gene sequences in DNA obtained from just a few cells, enabling them to identify different fingerprints and mutations quickly. Using microchip technology, it is now possible to analyze thousands of genes or DNA sequences simultaneously to get genetic profiles of individual cancers.

"You must follow the revolution in technology or you fall behind," said Dr. Cordon-Cardo, a technology buff who enjoys applying new technologies to clinical situations. "Cancers usually have hundreds of mutations, most of which are silent and some of which are very aggressive," he said. "The trick is to figure out what is really important among the background noise—which mutations allow a cell to escape the normal controls of proliferation and death and become immortal inside a mortal body."

But this is not work one researcher can do alone. "My life is a life of collaboration," he said. Among his colleagues are molecular biologists who identify genes and their mutations, crystallographers who study the structure of molecules, immunologists who devise ways of stimulating immune responses to cancer, developmental pharmacologists who engineer new drugs and clinicians who care for patients.

"The next step will be to make sure all these efforts by so many workers are not wasted," Dr. Cordon-Cardo said. "The discoveries in the laboratory must be brought into clinical practice to enhance the management of patients."

—JANE E. BRODY
April 1999

CLUES TO CANCER'S NEXT MOVE

After my surgery for a seemingly early breast cancer in February 1999, the pathologist who examined the tissue went beyond a microscopic confirmation of cancer and an examination of nearby lymph nodes to be sure they were free of errant cancer cells. In addition, my tumor was tested for at least three different molecular markers—genetically determined characteristics that would indicate how aggressive the cancer was and whether I might need chemotherapy in addition to radiation to increase my chances of a lasting cure. Based on these results and other physical characteristics of my cancer, I was spared the chemo ordeal.

Three months later, a friend had a similar experience but with a different outcome. Her breast cancer tested positive for a marker called HER-2/neu, an indication that it was highly aggressive and warranted aggressive treatment from the get-go, especially treatment with the drug Herceptin, specially designed to attack HER-2/neu tumors.

Herceptin is the first gene-based medication for cancer, but it most certainly won't be the last. And as more such medicines are developed, the chances of surviving cancers that were once uniformly fatal will continue to grow. For example, researchers at the University of Chicago Medical Center found that some women with small, early breast cancers and cancer-free lymph nodes carry a molecular marker called E-cadherin, which results in a survival rate half that of women whose tumors are free of this marker. Instead of 90 percent being alive and well years later, only 44 percent of women with the marker survived long-term. This finding suggests that aggressive postoperative chemotherapy may be needed to increase the survival rate of women with this marker.

Molecular markers for various cancers are being discovered at a dizzying rate. They are expected to enhance the ability to detect cancers at earlier, more curable stages and to enable oncologists to tailor-make treatments based on a tumor's likelihood of recurrence. In addition to those found in some breast cancers, genetically programmed molecular markers with treatment or diagnostic significance have been discovered for leukemia and cancers of the colon, prostate, kidney, esophagus and cervix, among others.

One molecular marker found in colon cancer cells may enable pathologists to detect the spread of cancer to lymph nodes that under the microscope appear to be cancer-free. At Thomas Jefferson Medical Center in Philadelphia, a recent study of colon cancer patients revealed that all who remained cancer-free six years after surgery had tested negative for a particular molecular marker in surrounding lymph nodes. But cancer recurred in all those who tested positive for this marker. By testing for this marker, oncologists can determine who needs treatments beyond surgical removal of the cancer.

The search is now on for markers that may be specific for cancers of the ovary, lung and pancreas, which usually produce no symptoms in their early stages and thus are not discovered until quite advanced and difficult to cure.

The formidable task ahead involves determining which of the many thousands of markers that characterize a particular tumor will yield the biggest therapeutic payoff. But experts have no doubt that before the end of the first decade of this new century, molecular profiling will revolutionize the treatment of many cancers.

—J.E.B.

P. Kirk Visscher

A Life Spent Among Bees Deciphering the Swarm

Scott Camazine *for* The New York Times

Partly shaded by the slender branches of a paloverde tree, his face just inches from a cluster of 5,000 honeybees, Dr. P. Kirk Visscher spoke into a walkie-talkie. "I think these bees are going to take off," he told a colleague stationed 160 yards west, at a wooden nest box that the bees had been scouting in their search for a new home.

Indeed, the cluster was becoming jumpier by the minute. Worker bees had begun "buzz running," plowing furiously through the masses of their sisters, as if prodding them to get ready. The buzzing grew louder as the entire cluster pulsated and changed shape.

"They're gonna go!" Dr. Visscher cried. His partner, Dr. Scott Camazine, came running to watch.

By dozens and then hundreds, the bees lifted off into the warm light of a February afternoon in the desert 150 miles east of Los Angeles. Within a minute, all 5,000 were airborne, zooming around in a circular holding pattern 20 or 30 feet in diameter, just barely above the heads of the two entomologists and Richard Vetter, a staff research associate from Dr. Visscher's laboratory at the University of California at Riverside. "Streaker" bees that had visited the nest box roared westward across the circle, signaling the others to head that way. A hum filled the dry air.

It was like standing in the midst of a tornado of bees. A visitor's first impulse might be to run for dear life or roll up in a ball on the ground, but it would surely give way to wonder at an event that even textbooks and scientific reports cannot resist describing as "spectacular."

"Very few people ever see this," Dr. Camazine said, smiling.

Such behavior, known as swarming, occurs when part of a bee colony leaves the nest to find a new home, and it has been described in detail by various scientists over the years. But important elements remain unexplained, and they are on a seemingly endless list of things that Dr. Visscher hopes to learn about bees.

An associate professor of entomology at Riverside, Dr. Visscher is both a honeybee biologist and a lifelong beekeeper. Of the 20,000 known species of bees, only about 6 make honey, and one of them, *Apis mellifera*, the best known and most common, is the focus of much of Dr. Visscher's research. He keeps 60 hives—about three million bees—on campus.

Dr. Visscher, 44, a large-framed, graying man more than six feet tall, has the wardrobe of a scientist who spends a lot of time hauling boxes of bees around in the field. His rumpled look and affable manner stand in contrast to his incisiveness on matters of biology.

Among his colleagues, Dr. Visscher is probably known best for the unusually broad range of his research interests, but that eclectic approach can leave him open to criticism that his efforts are too dispersed. "I do too many things," he said, but it would be hard for him to give up any of his disparate interests.

"He's an enormously creative scientist who's done research in a broad range of areas, and many are the critical areas of sociobiology," said Dr. Gene Robinson, who went to graduate school at Cornell with Dr. Visscher and is now an associate professor of entomology at the University of Illinois at Urbana-Champaign. "I suppose the flip side of this is that you can't say he's made his mark in this area or that. But his best papers are models of excellence."

Dr. Visscher studies the biology of bees themselves, but they also interest him as social insects with cooperative behaviors shaped by evolution. His scientific publications run from the lofty and theoretical to the unabashedly practical: he has pondered the evolutionary significance of egg laying by worker bees instead of the queen, helped monitor the advance of the notoriously fierce Africanized bees, or so-called killer bees, into California and studied ways to treat diseases in bees as well as the best soap to use to kill unwelcome swarms. He has even designed a Take-Out Trap, from a Chinese restaurant take-out carton, which is being used by California agencies to capture Africanized bees.

Last year, with Dr. Camazine and Mr. Vetter, he ventured into the world of medicine, letting himself be stung repeatedly in a study showing that contrary to traditional medical advice, the best thing to do for a bee sting is to yank out the stinger right away. The study may have struck entomologists as a bit whimsical, but it was carefully crafted and accepted by *The Lancet*, a respected medical journal.

It matters to Dr. Visscher that the insect he works with is commercially important. Beekeeping, including the sales of honey and bees and the rental of hives to pollinate crops, is a $200 million industry, and pollination by commercial colonies is estimated to add more than $9 billion a year to the value of crops in the United States. Although mite infestations have wiped out wild bee colonies in many parts of the nation, beekeepers have treated their hives successfully with chemicals, including fluvalinate and menthol. The mites may become resistant to the chemicals, however.

Bees have been part of Dr. Visscher's life for as long as he can remember. Growing up in Montana, where his father practiced medicine and kept bees as a hobby, he was working the hives and harvesting honey before he turned 6.

Even so, he did not always intend to become an entomologist. "It feels a little bit accidental," Dr. Visscher said. "I did always think I'd get a graduate degree. My father has a medical degree, and both grandfathers had Ph.D.'s. I considered medicine."

But when he went to Harvard, he found "a hotbed of social insect work," he said, and "that's when I really discovered entomology." The biology department was infused with excitement by the pioneering work of the sociobiologist Dr. Edward O. Wilson on ant societies. It captivated Dr. Visscher, and he declared a major in biology. He also kept bees on his back porch in Cambridge.

In 1978, when he had to decide on a topic for his senior thesis, bees were a natural. At the suggestion of one of his mentors, Dr. Thomas D. Seeley, he studied the bees that specialize in removing corpses from the nest and showed that they singled out the dead by smell. The work later led to a journal paper in 1983 called "The Honeybee Way of Death," in which Dr. Visscher called his subjects "undertaker bees," a designation that has stuck in the scientific literature ever since.

That senior thesis caught the eye of Dr. Roger A. Morse, a preeminent bee scientist at Cornell University, who invited Dr. Visscher there to do graduate work. During his years at Cornell, Dr. Visscher began to explore some of the ideas about bee societies that still inform his work today.

Among the work he has done that Dr. Visscher considers most important and intriguing are studies of what he calls "reproductive conflict." Although it is widely believed, and even stated in textbooks, that worker bees do not lay eggs in colonies with a queen, that is not the case. Some workers do; they lay unfertilized eggs that can develop into males, known as drones. In studies published in 1989, Dr. Visscher showed that workers laid far more eggs than researchers had previously thought. But hardly any of those eggs survive because other workers sniff them out and devour them.

The phenomenon raises evolutionary questions that he continues to study. If one accepts the idea that individual organisms evolve behaviors designed to pass along their genes, then it makes sense that the workers should try to reproduce. Their own offspring will, after all, carry more of their genes than the queen's offspring. Then again, if one worker can reproduce, so can her nestmates, most of whom are half-sisters—but their offspring carry even fewer of her genes than the queen's offspring. So it makes sense for the workers to behave as they do, destroying each other's eggs and attacking other workers whose ovaries begin to develop.

But why, then, do they lay eggs in the first place?

"It remains unresolved," Dr. Visscher said.

The project that he and Dr. Camazine are pursuing now concerns swarming. Bees generally swarm in spring, when the hive becomes too crowded. Workers begin rearing a new queen, and the old queen, often against her will, is swept out of the hive by a swarm that may contain 10,000 to 20,000 bees, between 30 percent and 70 percent of the original colony.

The swarm, clustered around the old queen, settles for a few days in an interim spot, often a tree branch, while scouts, arising from the ranks of worker bees, fly off to find a new site for a hive. When they return, the scouts do a dance for their nestmates that is similar to the one forager bees perform to direct others to food. Scouts make up perhaps 5 percent of the swarm. Some visit more than one site, and different scouts dance for different sites. Nonetheless, the moment arrives when the entire swarm takes to the air, and, somehow, all the bees go in the right direction to the new site.

"How do the bees achieve a unanimous decision?" Dr. Visscher asked.

Dr. Camazine said, "It's one of the last remaining great questions about honeybees, and Kirk is the ideal person to collaborate with." Dr. Camazine, an emergency room doctor, has put medicine aside to become an assistant professor of entomology at Pennsylvania State University.

Most researchers think that the scouts visit various sites, compare them somehow in whatever is the bee equivalent of a mind, pick the best one and then dance for it, so that much of the dancing eventually represents one site, which the swarm then occupies.

Dr. Visscher and Dr. Camazine are not so sure that bees make comparisons, and the experiments they have been conducting in Cactus City have been designed to test the idea. They take a small swarm from one of Dr. Visscher's hives and set it on a plywood stand in a stretch of desert that has virtually no sites, such as hollow trees or caves, that would attract bees looking for a home. That way, the swarm will go to nest boxes set up by the researchers, who post themselves at the boxes and mark scouts with a shade of paint that will reveal which box they visited.

Any bee that shows up at one box after visiting another—as shown by the paint marking—is destroyed. That eliminates the possibility of comparison by a single scout. If that kind of comparison is important, the scientists reason, then the swarm should be unable to decide where to go without it, or, at the very least, should take significantly longer to decide.

Although the experiments are still under way, Dr. Visscher said, there are hints that the findings will challenge existing notions about how swarms go about finding a new home.

Dr. Visscher expects the project to take several years and to open other avenues of inquiry. The idea that some scouts may dance for one site, visit another and then change their dance to favor the second site—something that in Dr. Visscher's earlier studies has been vanishingly rare—has inspired some researchers to suggest that bees might be capable of forming a mental construct based on what they have seen.

At this point, Dr. Visscher is not sure that is possible. "Sometimes it seems as if bees can do anything," he said. "Do they really act as automatons? Or is there some degree of awareness? It's difficult to think of an experiment to tell which an animal is doing. I wonder what it is to be a bee."

—DENISE GRADY
April 1997

SHE WHO DANCES LONGEST, WINS

So, how *do* swarms of bees pick a new home? Further experiments out in the desert at Cactus City confirmed what Dr. Visscher and Dr. Camazine had begun to suspect: despite what other researchers had thought, bees do not depend on their scouts to compare sites, judge one best and then dance for it. That would most likely require a thought process far beyond what goes on in the brain of the average bee.

Even when scientists eliminate the few scouts that have visited more than one site—and that could, theoretically, compare the sites—the colony still finds a new home, and takes no longer to do so than any other swarm.

But why do the bees follow one scout, and not another? The key to that mysterious process, Dr. Visscher said, seems to be the length of the dance. Just as a bee that is trying to tell her sisters about a great source of nectar will dance longer than one that has found a mediocre supply, a bee that has found a wonderful new neighborhood with a terrific home site will also dance longer.

"Our results seem to tell us that bees dance a larger number of circuits for sites they perceive to be better," Dr. Visscher said.

Other bees do not make a conscious decision to follow the longer dance; it is simply that a dance that goes on and on is more likely to catch their attention, whereas a short dance may go unnoticed, and the site it represents may be ignored. The dancing bees that are noticed recruit others to visit the site and dance for it, and if each dancer recruits several more, the votes in favor of a particular site can grow geometrically.

"It points in the direction of the bees not making a highly conscious decision between sites," Dr. Visscher said.

In February 1999, he and Dr. Camazine published their findings in the journal *Nature*.

But questions remain. At what point does the swarm "know" that a decision has been made, and a new home selected? And is it at that moment that the swarm departs for its new home, or is takeoff a separate decision?

"I think that will be our next big thrust," Dr. Visscher said.

He and his colleagues have intensified their efforts in the desert, now living out at the Cactus City site in a "rather creaky old RV" for days at a time, instead of making the 200-mile round trip every day.

They have begun to suspect that buzz running has something to do with takeoff, and they are studying it closely. They have already begun to realize that it occurs earlier in the swarming process than they had previously thought.

Dr. Visscher explained: "It may be playing some sort of role in the bees' assessing, 'Have enough scouts scouted the site now for us to be sure that a lot of recruitment has gone on and it's a good site?' "

In the meantime, he and his research partners have continued their eclectic pursuits, studying the properties of various bee and wasp venoms, and tracking the migration of Africanized bees in California, where, in September 1999, an 83-year-old man in Long Beach became the first person in the state to be stung to death by the aggressive insects.

—D.G.

Paul C. Sereno

Imp's Evolution to Fossil Finder

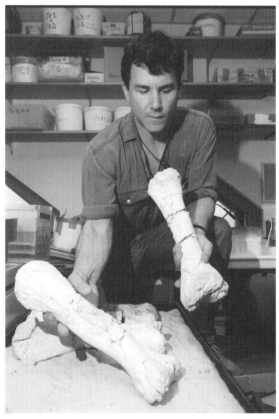

Steve Kagan *for* The New York Times

It is 9 A.M. on a weekday, and Susan Lesher's fifth-grade class from the University of Chicago Laboratory School is agitatedly awaiting the entrance of the dinosaur hunter Dr. Paul C. Sereno. Minutes later, the door to the lecture room at the university swings open and Dr. Sereno—in jeans, a West African charm necklace and one turquoise earring—strides to the front, grinning broadly at his adulatory audience. Jaws fall open, revealing braces on teeth, and a few hands shoot up with burning questions, prompting a raised-eyebrow caution from Ms. Lesher.

"Griffin, please wait, please wait," she whispers gently from halfway across the room.

The man needs no introduction to these dinosaur disciples. Just the other week, he and his team at the University of Chicago announced their discovery of fossil remains of two large predatory dinosaurs that must have been the terror of the flood plain they inhabited 90 million years ago: *Carcharodontosaurus saharicus,* which means "shark-toothed reptile from the Sahara," a bipedal dinosaur bigger than *Tyrannosaurus rex,* and *Deltadromeus agilis,* for "agile delta runner," a relatively lithe, fleet creature. Carcharodontosaurus was identified earlier in this century, but Deltadromeus is a new species.

The pupils have seen Dr. Sereno's rugged image on the morning news, pored over his make-or-break expeditions to Niger and Morocco, and prepared for this visit to his laboratory by studying a curriculum he wrote with Gabrielle Lyon, a University of Chicago graduate who participated in both expeditions. For the next 90 minutes, Dr. Sereno and Ms. Lyon, who spotted the glint of the crystallized Deltadromeus marrow under the glaring Moroccan Sun, regale the pupils with facts and stories about their work and the lives of dinosaurs.

In recent years, Dr. Sereno's time has been increasingly in demand, but visits by schoolchildren frequently dot his agenda. "This could be the experience that changes a kid's life," he said after the session. Dr. Sereno's account of his own uneven childhood in nearby Naperville, Illinois, suggests a reason for his belief in turning points.

He was considered a poor student. "I wasn't reading in second grade," he said. "I couldn't tell time in third grade, and I nearly flunked sixth grade."

He was also incorrigibly mischievous and accident-prone. He tried derailing trains and pelting school windows with rocks, and he landed in several body-mutilating accidents involving knives and bikes. Twice he was rendered unconscious in gym class mishaps.

It was another misdemeanor that finally set him on a more virtuous course. "Just a minute. I still have it," he said, rushing from the room and returning a moment later with the 1956 edition of "The Fossil Book," wrapped in yellowed cellophane and stamped on all sides with the words *Naperville Community High School Library.*

Yes, he stole it. "There were two copies," said Dr. Sereno, who at 38 views the prank a bit sheepishly. It is a book packed with photographs of fossils, slimy creatures and colorful stones, and the illustrations captivated him at a time when he was discovering his own talent in painting. With the unwavering support of his parents (whose five other children now also have Ph.D.'s), he enrolled at Northern Illinois University in De Kalb, majoring in biology and studio art. Then, on a trip to the American Museum of Natural History in New York, Dr. Sereno became fascinated with the stories of paleontological expeditions.

"I could combine art, travel, science, adventure, biology, paleontology and geology," he remembered thinking. "Right then, I knew exactly what I wanted to be."

From there, Dr. Sereno went on to earn a Ph.D. in geology at Columbia University in New York. He got a job at Chicago, where he has taught since 1987 in the department of organismal biology and anatomy.

In the last decade, he has led and participated in expeditions to South America, China and, most recently, Africa, where his team made the first relatively complete discovery of dinosaur remains from the Late Cretaceous period, the final stage of dinosaur evolution. But in a field where, according to one paleontologist's estimate, a new dinosaur species is discovered on the average of every seven weeks, it is not only Dr. Sereno's discoveries that impress some of his colleagues.

"Sereno's accomplishments aren't what he gets credit for in the popular media," said Dr. David E. Fastovsky, a paleontologist at the University of Rhode Island in Kingston and co-author with Dr. David B. Weishampel of *The Evolution and Extinction of the Dinosaurs* (Cambridge University Press, 1996). "He has a truly well-developed, high-quality eye and breadth of knowledge of the character of dinosaurs. He knows which dinosaur is related to which as well as anybody out there."

Dr. Sereno's main interest is in explaining the genealogy of dinosaurs using cladistic analysis, a tool that presents dinosaur relationships as a hierarchy of shared characteristics. In a cladogram, each branching point, or node, marks a hypothesized evolutionary step between organisms. For example, the earliest branching point in dinosaur evolution was between the Ornithischia, whose defining characteristics include a backward-rotated pelvis, and Saurischia, whose defining characteristics include an asymmetrical hand. By contrast, the old family trees postulated relationships between organisms without presenting rigorous anatomical evidence for such connections, Dr. Sereno said.

In Dr. Sereno's African expeditions, he sought to learn how dinosaurs had evolved once the supercontinent known as Pangaea began to break in two around 150 million years ago. One landmass, which would later separate into the African and South American continents, moved south. Another landmass moved north. If the break had been final and clean, as had been hypothesized, then one would expect to see a clear branching point between northern and southern dinosaur lines as well.

But the new Moroccan finds appear to tell a different story. Both the Carcharodontosaurus skull and the Deltadromeus skeleton were discovered in sediments that date from tens of millions of years after the start of the continental split. But both of these dinosaurs, from the southern landmass, are most closely related to dinosaurs on the northern landmass. Dr. Sereno said that might mean there was a land connection between the two continents for much longer than has been believed.

Dr. Sereno came upon the first softball-size fragment of the Carcharodontosaurus skull while surveying the side of a 500-foot-high incline. When he saw the fragment, he froze in his tracks to avoid crushing any other bits that might have been underfoot. Then he carefully picked up the fragment.

He immediately saw a distinctive triangular depression that is a distinguishing characteristic of theropods, a group of meat-eating dinosaurs like *Tyrannosaurus rex.*

"My eyes just about popped out of my head," he said. "You pick up so many pieces like this that are so weathered. You're usually a million years too late."

But in this case, Dr. Sereno could see that the break was fresh. He guessed that the skull fragment had fallen off recently—maybe in the last 20 years—from higher up the cliff. After several minutes of searching, he recognized the palm-sized matching end of the piece protruding from a spot 20 feet up.

Over the next five days, all nine students and paleontologists worked at the site, painstakingly clearing away the soft rock around the animal's braincase and snout. Since much of the fossil headed straight into the cliff, the project was akin to removing one brick from a wall without having the part above collapse.

"You have to make a lot of Michelangelo-like decisions about what the rock can take," Dr. Sereno said.

His solution was to get rid of the cliff above the fossil. "We moved tons of rock from the side of the cliff to get it out," Dr. Sereno said. As the team members worked, he said, one graduate student was assigned to cover the exposed fossil with his body.

"He was like this down there," Dr. Sereno said, flexing his knees slightly, looking up, squinting and holding up his bent arms above his head. "When a boulder came, he just knocked it away."

He was only half joking. The physical demands on the eight men and one woman on the team were extreme. They were in the Sahara Desert in the middle of summer. Some team members lost 20 pounds in 60 days in the field.

Backbreaking labor was interspersed with exacting brush and pick work. The condition of the Carcharodontosaurus bone was like that of a shattered porcelain plate. To remove the fossil in one piece, the team hardened it with a glue-like substance as it was unearthed, then wrapped it tightly in paper, plaster and burlap. Once removed, the brain case and snout weighed more than 300 pounds, much more than when the animal was alive because minerals in fossilized bone are heavier than the bone they have replaced and the cavities inside are packed with other deposits. In the broken rock below the find, the team members worked doggedly to recover 400 smaller pieces of the skull.

"We would not have a skull if we had not sifted the side of that cliff," Dr. Sereno said.

Perhaps because of his background in art, Dr. Sereno is keenly involved in the visual presentation of his work. He works closely with his laboratory's illustrator, Carol Abraczinskas, and shot his own video of the Niger expedition. He and his students helped make the cast of the Carcharodontosaurus skull that is already on display at the university's John Crerar Library. On a brief visit there, he fretted momentarily over whether the lighting was good enough to show the unusual grooves in the monster's five-inch-long teeth.

Back in Dr. Sereno's lab, work on the Morocco material has been continuing day and sometimes night, as he, graduate students and visiting scholars sort through three tons of material from the expedition, including several canvas bags of sediment.

"If we pull out five or six mammal crowns out of several hundred pounds of that sediment, it would be hot stuff," Dr. Sereno said, noting that 90 million years ago, mammals had been no larger than mice.

Once everything is catalogued and studied, it will eventually return to Rabat in Morocco, where some fossils will be displayed at the Museum of the Geological Survey. To look for fossils in the country, Dr. Sereno had to get permission from the Moroccan Government, which is losing a battle against unauthorized dinosaur fossil hunters who smuggle the precious remains out of the country and sell them on the antiquities market. In the region in which Dr. Sereno and his team were working, they came upon hundreds of holes showing where fossils had been extracted.

That these treasures are disappearing into the cabinets of collectors at a rate far greater than it takes for the erosion process to expose the fossils frustrates and saddens Dr. Sereno, he said, and partly explains his rush to visit such sites around the world. He will do some scouting this summer, and the team will head to northern Patagonia in Argentina next November.

"We're going to be exploring and publishing as fast as we can for a number of years," he said.

—Brenda Fowler
June 1996

INDIANA SERENO

The six hundred guests hushed as the evening's most spectacular prize went on the auction block: an all-expenses-paid trip for two to visit Paul Sereno's excavation site during his next fossil-hunting expedition to the Sahara of Niger.

Starting bid: $20,000.

Beneficiary: Project Exploration, an educational organization Dr. Sereno and his wife, Gabrielle Lyon, founded in 1998 to teach young people about paleontology.

The first bids rushed in from the lavishly decorated tables in the ballroom at Navy Pier in Chicago, but had apparently stalled at $23,000 when Dr. Sereno himself, in a sleek black suit and T-shirt, leaned into the mike.

"You will be met at the airport, escorted to really what is the center of Tuareg culture," he said, as images of sunburned and dusty excavators crouching over fossils flashed on the three screens behind him. "From there you will travel out to some of the most beautiful desert scenes in the Ténéré to partake in digging up . . . the next discovery."

Suddenly, a messenger arrived at the stage and handed up a note.

"We have a bid on the floor of forty thousand dollars," boomed Bill Kurtis, the fund-raiser's emcee and the anchor of several A&E network documentaries. Flashing the grin that must have helped rank him as one of *People* magazine's 50 Most Beautiful People in 1997, Dr. Sereno buckled slightly in disbelief.

The unexpectedly high bid illustrates not only Dr. Sereno's ever soaring popularity as an explorer and guide to the world of paleontology, but also his ingenuity in developing new ways to market himself and his doings.

In all, the December 1999 fund-raiser brought in about $130,000 for Project Exploration, which the previous summer sent more than a dozen teenagers on an excavation to Montana. For Dr. Sereno, the highlight was the unveiling—amid a swarm of spotlights and the live beat of African drums—of his team's latest discovery, a sauropod named *Jobaria tiguidensis*. The precise relevance to paleontology of this enormous herbivore that lived in Africa 135 million years ago surely escaped the vast majority of the people in the audience; their gaze was instead on Dr. Sereno, whose adventures, including expeditions to South America and Africa, have been profiled in no fewer than eight television documentaries in the last eight years.

"To hear him speak, would you think he's a scientist?" asked Connie Rivera, chief executive of the American Dietetic Association and a volunteer organizer for the event who, like many people there, knew Dr. Sereno personally. "There's nothing nerdy about Paul. Everything is graphically presented."

Indeed, among the exhibits on display at the fund-raiser was the actual canvas tent housing the actual cots and the actual plastic mugs used by Dr. Sereno and Ms. Lyon during the 1997 Niger expedition.

"We wanted it to be the party of the year, and I think we succeeded," said Dr. Sereno, who clearly enjoyed mingling with his admiring guests.

Later, over breakfast in a basement cafeteria at the University of Chicago, Dr. Sereno outlined a dozen more ideas for projects, some well under way, some still just dreams, for expeditions, events, a popular book and other possibilities to introduce people to dinosaurs and general science.

"We want to establish a home that might center around a herd of dinosaurs," he said. "I don't think in terms of a normal museum."

In some museums, curators who themselves never saw the fossils in the ground are faced with building an animal no one has ever seen before. Dr. Sereno's team, by contrast, takes the fossils from the ground through the lab and onto display. Last year he opened his own lab, staffed partly with volunteers, in which he can prepare fossils and mold and cast replicas of the skeletons for mounting. He said he starts thinking about how to display an animal even before it is out of the ground, in some cases even deciding what to dig based on whether he will be able to reconstruct it.

Despite Dr. Sereno's confidence before the public, he is aware of mutterings among colleagues who disdain his flashy approach to the discipline.

"If there's publicity, it looks as if you seek publicity, and that seems like an unscientific thing," he explained, suggesting that it is not. "When we announce something, we try to get a *Science* article behind it."

—B.F.

Terry DeBruyn

Black Bears Up Close and Personal

David Binder/The New York Times

Snow fell steadily. It was already 20 inches deep in Hiawatha National Forest, Michigan, when Terry DeBruyn scrabbled barehanded in numbing cold to open up the entrance to the den of a black bear. Soon he was slithering into the dark lair armed only with a flashlight.

His quarry, her powerful jaws inches from his face, was a 5-year-old female and her yearlings. Not to worry, they were in semihibernation, eyes open, but not stirring much. Even fully awake, black bears are much more likely to flee than to attack humans.

Using a stab stick tipped with a hypodermic needle, he sedated the adult bear, named Priscilla, and her two male yearlings. Helped by two other men, Mr. DeBruyn, a wildlife biologist, pulled the bears out of the den and onto plastic bedrolls. The silence was broken by the keening of the north wind in the pines and the occasional cheeping of a flock of siskins.

The year before, in 1994, Priscilla had shed a collar carrying a tiny radio transmitter, but one of her yearlings still bore a transmitter. That enabled Mr. DeBruyn to locate the family in the den under a fallen white cedar in the Hiawatha National Forest in Alger County, Michigan.

Together with Dr. Larry Visser of the Michigan Department of Natural Resources and Kevin Doran of the United States Forest Service, he made the prostrate bears as comfortable as possible with blankets and with his red-and-black-checked jacket. They set to work to measure claws, teeth, girths, heads, paws, lengths and weights, and Priscilla's teats. Blood samples were taken; antibiotics were injected. Finally, Mr. DeBruyn slid the bears back into the den and covered up the entrance with pine branches and snow. The operation, including trekking on snowshoes, took about two hours.

If Terry DeBruyn had an Indian name it would probably be "Walks With Bears."

He pads through the forests of the Upper Peninsula of Michigan close to American black bears on a quest, as he put it, "to learn how a bear views the landscape." Two of the adult bears have become accustomed, or habituated, to his company.

Wildlife biologists have been investigating the black bears (*Ursus americanus*) for decades in most of the 28 states where they are indigenous, but, unlike Mr. DeBruyn, few have managed to habituate themselves with wild bears. None has spent as much time with individual habituated animals.

"Over 1,000 hours walking with bears," he said, sometimes as long as 15 hours from dawn until after dusk.

Calling upon the skills of a woodsman, traditional scientific methods and space-age technology, he has mapped the meanderings of bears over long periods, learned new details about what they eat, and documented 36 different bear activities down to minutes and seconds.

After the encounter with Priscilla and her yearlings, Mr. DeBruyn and his group packed up their equipment, including a shovel, handsaw, medical kit, radio receiver, ropes, scales, tape measure, calipers and netting, and set off on snowshoes to the next female bear, named Lil. The problem was Lil's female yearling, who defied Mr. DeBruyn's efforts to snare her, suddenly shooting through the roof of the den in an explosion of branches and snow and scooting away from pursuers floundering knee-deep in fluffy white crystals.

"Score four for the biologists and one for the bears," Dr. Visser said as darkness fell.

Mr. DeBruyn and his team repeated their annual den calls for three more days as part of his research for a doctorate in wildlife biology at Michigan Technological University in Houghton. In a broader sense the research could yield valuable information for the management of bears and forest habitats as healthy components of the environment. In the case of bears, a key question is, how large do populations need to be to permit hunting without becoming nuisances to human beings?

Bear hunting has become a contentious issue in several states. Colorado and Oregon have banned bait hunting of bears, and Michigan faces a highly emotional referendum in 1996 on whether to ban not only baiting but also the hunting of bears with dogs.

The Michigan Department of Natural Resources is assisting Mr. DeBruyn's research because it will help the state calibrate the number of bears to be hunted. For the Forest Service, which is also contributing funds and research assistance, the management of bear habitats involves such issues as clear-cutting timber. Hunting organizations also provide funds for his research.

The range for Mr. DeBruyn's hikes, roughly 100 square miles, is mostly second-growth, sometimes third-growth timber—mixed stands of red and sugar maples, beeches, birches, aspens, hemlocks, cedars, black spruces, red pines, jack pines and white pines. The area is dotted with lakes, swamps and meadows and crossed by creeks. It has patches of raspberries, blueberries and wild cherries, and decaying stumps and logs rich in ant colonies and other insect pupae. That is, good bear country.

Along the way Mr. DeBruyn has become a kind of apostle for bears. In his role as educator he preaches his message in public forums, in schools and on television. One day during the latest week of den work he invited 14 schoolchildren to observe the den of a bear called J. C. and her four cubs. On another day he invited two local bear hunters to snowshoe to a high ridge amid pines and hardwoods to see 135-pound Vicky and her two male cubs. He befriended one of the hunters last year when the man telephoned to say that a collared bear had come into his gun sights and that he had spared her. It was one of Mr. DeBruyn's habituated bears.

Following an exhausting day at the dens he conveyed the sum of his experience to the Sierra Club chapter in Marquette, Michigan: "Bears can get along with us—the real question is whether we can get along with bears."

Initially, bears were an avocation for Mr. DeBruyn. "I hunted bear in 1981 and then off and on until 1989," he said. "I shot six with a bow and arrow. Then I started to ask questions: 'Am I doing harm if I kill a sow?' 'How many bears are there?' 'How big is their home range?' I didn't get very good answers." He was a soil conservationist at the time, earning $40,000 a year, with a house on a lake.

One day, he drove to Minnesota to meet Dr. Lynn L. Rogers, a former United States Forest Service biologist, who then introduced him to a habituated bear. Mr. DeBruyn became a convert. On the way back he laid out a plan for a study on the Upper Peninsula, where 90 percent of Michigan's 8,000 to 10,000 bears live. He quit soil conservation and earned a master's degree in wildlife biology at Northern Michigan University. He is continuing the study on an annual budget of $50,000 from 10 sources. Of this his personal income is $12,000.

He employs traditional scientific methods, collecting hundreds of scats (bear droppings)—three freezers full at one point—to examine minutely the evidence of what bears consume. But he also uses global positioning satellites with a computer he totes in a yellow knapsack.

With the computer's small dish receiver and antenna he can punch in bear data from any site in the woods and fix a location by way of a Navstar satellite accurate within five yards for various activities. He types NU for nursing, FB for feeding on berries, RR for rest recumbent, FA for feeding on ants, WM for wandering/meandering and any of a dozen other actions. The accumulated data enable Mr. DeBruyn to map the travels and actions of a bear over long periods, matching them to Landsat images that show the types of forest cover as well as swamps and lakes in his study region. For this he uses a "bear watch" computer program developed by Dr. Rogers. Using his knowledge of the soils in the area—loamy in the western half, sandy in the east—he has been able to document the bear habitat in minute detail.

"Anybody who habituates a bear is going to get some excellent data," said Dr. Rogers. "I think only Terry and I have done it."

Mr. DeBruyn listed his first habituated bear as No. 9. But because of the shuffling dancelike steps she sometimes took, in "a stiff leg walk with toes splayed," he said, he began calling her Carmen. He surmised that this might be an "expression of tension or territorial marking behavior."

The relationship began in June 1990 after he caught her in a barrel trap—two 50-gallon drums welded together—with bait tied to a trigger attached to a door. After briefly tranquilizing her, he attached a stout leather collar with a small radio transmitter around her neck.

Carmen was then 4 years old and weighed about 120 pounds. Mr. DeBruyn subsequently tracked her using telemetry with a small receiver with earphones and an H-shaped antenna—the closer the bear, the louder the signal.

In the spring of 1991 he repeatedly came upon Carmen and her three cubs, who scampered up a tree in typical bear behavior, "enabling me to approach the refuge tree, talk and leave food," he said. He left doughnuts, which Carmen consumed while remaining suspicious.

But, he said, "after approximately 150 hours of being approached, talked to and fed (day-old leftovers from Mister Donut in Marquette), the bear and her cubs tolerated observation from distances of 2 to 12 meters as she foraged, nursed her cubs, napped."

With that his bear walks commenced. He trailed Carmen and the cubs periodically, logging 100 hours with them until they denned in the late autumn. In the winter snow, he located the mother with her yearlings again, pulled them out of their den, sedating Carmen so that she and her offspring could be weighed and measured and blood samples taken. The cubs were also tagged.

He repeated this procedure with Carmen year after year, eventually abandoning the lure of doughnuts and logging 700 hours with her. Ultimately, he was able to approach Carmen more or less at will, observing her communicating with her cubs and yearlings and recognizing some of her vocal signals. "She clicks her teeth to send them up a tree, and blows, 'Uff,' to bring them down," he said. "Clicking her jaws hard means, 'I'm here.' When she pops her lips it means you did something wrong. Slapping the ground with a paw means you're pushing her. Then 'Hawhawhaw' is the top expression of anger."

One day he got a lip popping from Carmen when he deliberately placed a doughnut on a log next to a harmless pine snake. Then she marched off disdainfully. In circumstances like this, he communicates with her in a soft voice, saying repeatedly, "It's OK," as he approaches. "Then she calms," he said.

In 1992 he caught up with one of Carmen's yearlings and was able to repeat the habituation process. He named her Nettie, after his wife,

Annette. "It was an intimate gesture," Mrs. DeBruyn said. For his birthday that year she gave him a vanity license plate that says URSUS.

From the walks he has learned new information about what bears eat—for instance, a heavy diet of herbs from wetlands in the spring such as water plantain, sarsaparilla, jack-in-the-pulpit, ferns, thistles, cattails and water parsnips; berries and ants in the summer; berries and nuts in the autumn. Black bears are omnivorous, Mr. DeBruyn noted, although his studies show that meat—a fawn for example—constitutes about 5 percent of their diet, while the consumption of ants might be double that in his area.

He has also charted the often overlapping ranges of the collared bears, determining that his females, bound to nursing cubs or yearlings, tended to stay in an area of less than six square miles, while males, unfettered by any family obligations, might range 50 miles in a day in search of food or a female. "With males it could be breakfast in Munising and dinner in Sault Sainte Marie," Mr. DeBruyn said.

On his walks Mr. DeBruyn has also been able to document 36 different bear activities over a period of 100 daylight hours, among them sniffing, lying on a side, playing with cubs, grooming, marking territory, swimming, feeding on hornets, chasing other bears and yawning.

He has had a few brushes with his bears, all of them benign, perhaps the worst being when one of Vicky's cubs gashed his left cheek with a half-inch claw as he was putting him back in the den. Bears also flattened the two rear tires on his van by biting them while trying to get at bait inside— "a tough one to explain to the insurance company," he said.

Mr. DeBruyn expects to complete a Ph.D. later this year, but he would like to continue his studies for another five years, he said, "to gain a little more understanding of what bears are about." To this end he wants to attach newly developed global positioning satellite collars on some of his bears "to get a macro view of bear movements." But such collars are costly.

In the meantime, he had the satisfaction of finding Carmen in her den with three new cubs in March and weighing a healthy 205 pounds.

—DAVID BINDER
April 1995

BLACK BEARS AND GRIZZLIES, OH MY

Terry DeBruyn, his doctorate now earned, doesn't spend quite as much time with bears as he did five years ago in the Upper Peninsula of Michigan, where by

dint of great patience and a radio receiver he logged several thousand hours on a friendly footing with half a dozen members of the *Ursus americanus* species.

He is in Alaska now working not only with black bears but also with Alaskan brown bears and grizzlies, which are not as approachable as *Ursus americanus*.

Dr. DeBruyn's familiarization with black bears—technically it is called habituation—has led him toward broader conclusions about relations between humans and wildlife.

"For me it is a given that bears depend on the availability of nutritious food in the spring," he said, "which means tiny little wetlands in the forests. Bears pick out plants that have 3 to 4 percent nitrogen, which translates into 20 percent protein, like sweet cicely in the parsley family. They follow nitrogen pathways in the forest, selecting these plants. I see the little wetlands disappearing because timber cutters go right up to the edge and they dry up, letting other plants encroach. The bottom line is do you do anything about how we manage wetlands in forests that support a huge variety of wildlife—snakes, birds, small mammals."

After completing his doctorate at Michigan Technological University in Houghton in 1997, Dr. DeBruyn moved to Florida, where he worked for 18 months as a bear biologist for the state's Game and Freshwater Fish Commission. This shifted him from a region with a population of 310,518 people and upward of 7,500 bears to a state with 17 million people and fewer than 1,500 bears. He also worked on the manuscript of his book, *Walking With Bears,* but found it impossible "to write about den work while sitting on a deck with temperatures at 90 degrees."

So he returned to the Upper Peninsula as an adjunct professor at the university in Houghton and resumed work with his black bears. But the situation had changed in his absence. A hunter with a rifle killed Carmen, the first bear he had habituated, despite her radio collar. In the autumn of 1998 he discovered that an archer had shot Carmen's female radio-collared offspring, Nettie. Only one of Carmen's cubs survives. His collared bear Priscilla was killed by a hunter with hounds.

He also learned that some hunters in the range he was working "resented the project because I had put a crimp in their style." These scofflaws, as he called them, had taken revenge by killing collared bears and more. "There was some harassment, screwing with my car, putting nails in the road," he said. He was saddened because "for a long time I worked pretty hard to get along with the hunters."

He closed down his Upper Peninsula bear project and finished *Walking With Bears,* which was published in November 1999. Finding he couldn't make ends meet on modest grants for bear research and his adjunct professorship, he found a temporary job at Katmai National Park in Alaska.

There he worked with brown bears along the Brooks River, where they feed on salmon. Compared to black bears, he said: "They have an entirely different disposition, a different personality. They are better fed, even satiated. They don't flee like black bears."

Grizzlies, while cousins of the brown bears, are "interior bears who tend to be more aggressive because they need food."

In November 1999 he was appointed as a wildlife biologist for all of Alaska with the Department of Interior's National Park Service, based in Anchorage.

As one of only two American biologists who have managed to get black bears habituated to humans in the wild—the other being Lynn Rogers, director of the North American Bear Center in Ely, Minnesota—Dr. DeBruyn said he would not try that with the big Alaskan bears "because of the way they have evolved—they're ornery."

Why bother with bears? In his book, Dr. DeBruyn says they can be viewed as "umbrella species" whose preservation guarantees the survival of many other organisms, or as "keystone species" whose extinction would signify the end of an entire ecosystem, or even, through studies of bone physiology in hibernation, as a means to understand medical problems like osteoporosis.

But he prefers to think on the "intrinsic worth in just knowing they're out there."

—D.B.

Benjamin D. Santer

Blaming Humans for a Warmer World

Darcy Padilla for The New York Times

D r. Benjamin D. Santer, a shy, even-spoken, 41-year-old American climatologist who climbs mountains, runs marathons and enjoys a reputation for careful and scrupulous work, is the chief author of what may be the most important finding of the decade in atmospheric science: that human activity is probably causing some measure of global climate change, as environmentalists have long assumed and skeptics have long denied.

The finding, issued for the first time in December 1995 by a panel of scientists meeting under United Nations sponsorship in Madrid, left open the question of just how large the human impact on climate is. The question is perhaps the hottest and most urgent in climatology today.

Dr. Santer is in the forefront of a rapidly unfolding effort to answer it, and he has been itching to get back to the chase now that he has finished his stint as main author of the crucial Chapter 8 of the United Nations report, which details where the quest has led so far.

Ordinarily he is so absorbed in the work of analyzing the effects of heat-trapping carbon dioxide and other industrial emissions on climate that he has programmed his computer here in a small, spare room at the Lawrence Livermore National Laboratory in Livermore, California, to sound a cuckoo clock alarm every hour. Otherwise, he says, his absorption becomes "catatonic."

Lately, though, he has been able to accomplish little. His role on the UN panel, which he did not seek, placed him squarely in the sometimes dangerous intersection of science and politics, provoking so much contention that he has had scarcely a moment to deal with anything else.

Instead of plunging back into his research, he has been forced to engage in public combat with a tiny but vocal group of industry lobbyists and contrarian scientists, including Dr. Frederick Seitz, a former president of the National Academy of Sciences and of Rockefeller University. They charge that unauthorized and politically inspired changes were made in Chapter 8 after it was approved in Madrid, and that the changes served to underplay uncertainties about the effects of human activities on climate.

Dr. Santer "must presumably take the major responsibility" for "a disturbing corruption of the peer review process," Dr. Seitz wrote in an op-ed article in *The Wall Street Journal*.

Dr. Santer has struck back hard, like the trim middleweight he resembles, throwing verbal jabs and crosses. He and he alone did indeed alter Chapter 8 after the Madrid meeting, he said, because the scientists gathered there accepted the chapter, after long discussion, only on the condition that he do so. The result, he maintains, is a clearer and more accurate statement of the relevant science than the earlier draft. And he points out that the revised chapter devotes a long and detailed section to uncertainties.

The post-Madrid revisions left unchanged the chapter's basic conclusion that the scientific evidence so far points toward a human influence on climate. The chapter, which did not require line-by-line approval, provided the underpinning for the Madrid group's official finding—formally approved word by word—that "the balance of evidence suggests a discernible human influence on global climate." That conclusion, contained in a separate summary for policy makers, also remains unchanged.

The panel advises the world's governments, which are negotiating reductions in greenhouse emissions.

Dr. Santer says that neither his industry critics nor Dr. Seitz contacted him or anyone else associated with Chapter 8 for their side of the story, and that by failing to do so, Dr. Seitz "failed to behave as a responsible scientist should." Further, he says, Dr. Seitz, a nuclear physicist who is not a climate expert, is unqualified to judge the science of the case. And he says that his critics have done no peer-reviewed research on the question; that they, not he, are the ones subverting the science and that their charges "would be ludicrous if they were not so serious."

Top officials of the United Nations panel have strongly backed Dr. Santer, calling the accusations "rubbish" and saying that the alterations were faithful to the mandate of the Madrid group. Most governments, including that of the United States, have endorsed the revised report. And many climate experts have flocked to Dr. Santer's defense.

The criticism "has actually rallied a lot of the serious scientists around him," said Dr. John F. B. Mitchell, a British climatologist who was Dr. Santer's Ph.D. examiner a decade ago

The critics have not relented. In Congress, conservative Republican allies of the critics have also weighed in, raising questions about the role of the Department of Energy, which has contracts with Lawrence Livermore, in financing the work of chief authors of the report, including Dr. Santer. This has raised fears among climate scientists of political intimidation.

As for himself, Dr. Santer says, "I've been taken out of science" by having to deal with the attack.

Dr. Santer is widely regarded by colleagues as being among the most careful and thorough of scientists and the straightest of straight arrows, and as being apolitical. "I have no politics," he said. Dr. Karl Taylor, a colleague at Livermore who has worked with him closely, says that Dr. Santer "has set very high scientific and ethical standards for himself."

Dr. Mitchell said it was "quite ironic that somebody like Ben should be accused of things like this, because I don't think you could find anyone less likely to do that sort of thing."

One of the chief critics says no personal attack on Dr. Santer was intended. "There wasn't any implication that we thought he was doing anything wrong," said John Shlaes, director of the Global Climate Coalition, an industry lobbying group that first raised the questions about Chapter 8. Mr. Shlaes complained that the post-Madrid changes in the chapter were not publicly revealed, discussed or formally approved before the report was published. "It was basically the process we were addressing," he said.

Dr. Santer, like his forebears, has been no stranger to the character tests imposed by adversity. Two of his grandparents died in a Nazi concentration camp. His Jewish father fled Nazi Germany and migrated to the United States, where he joined the American Army. He later landed and was wounded at Normandy, served as an interpreter at the Nuremberg war-crimes trials, met and married his wife in Munich and then became a businessman in Washington, where young Ben was born in 1955. He attended the public schools of Bethesda, Maryland, moving to Germany at the age of 10 when his father was transferred there.

In Germany, he went to a school for the children of British Army personnel, where, on the first day, the English teacher grabbed his exercise book "and whacked it around my head, and said, 'Santer, we're in a British school, and you'll write in pen, not in pencil.' "

"In the end," Dr. Santer said, "I got an excellent education." He graduated with top honors in 1976 from the University of East Anglia in Britain with a degree in environmental sciences.

To his dismay, his British education availed him little in the job market when he returned to his parents' home, then in the Baltimore area. He bounced around for the next few years, working at various times as a soccer teacher, a German teacher for Berlitz and an assembler in a zipper factory, at which point, he says, he found himself "down and out in Seattle." He made two stabs at a doctorate at East Anglia, abandoning both.

About that same time he nearly died after falling into a crevasse in the French Alps, where some friends had invited him on a mountain-climbing holiday to take his mind off his troubles. He painstakingly worked his way toward "a little blue slit of sky" at the top of the crevasse, and the experience, he says, has "made me realize that there are things much more serious than the present controversy over Chapter 8."

He soon made a third attempt to earn a doctorate at East Anglia, which boasts one of the world's top climatology departments, and this time he succeeded.

"I found it fascinating," he said, "the idea that humans could have a potentially large impact on climate." In his dissertation, Dr. Santer used statistical techniques to investigate the accuracy with which computerized models of the climate system simulated regional climates.

He soon moved to another leading climatological laboratory, the Max Planck Institute for Meteorology in Hamburg, where he worked for the first time on the problem of detecting the signal of human-caused climate change, especially global warming—the "greenhouse fingerprint." He also met his wife, Heike, during the Hamburg stint, and they now have a 3-year-old son, Nicholas.

Since moving to Livermore in 1992, Dr. Santer has grappled with the related problems of testing the validity of climate models and searching for the greenhouse fingerprint.

Considerable progress has been made on the fingerprint issue in the last five years. Until recently, most efforts to detect a human impact on climate focused on the average global temperature, which has risen by about 1 degree Fahrenheit in the last century. But it proved too difficult to tell whether this small amount of planetwide average warming was the result of human activity or of the climate's natural fluctuations. So a small group of researchers, including Dr. Santer, turned to a different strategy.

Behind the rising average global temperature lies a widely varying spatial pattern: some parts of the world have warmed more than others, and some have actually cooled. The idea of the new strategy was to examine observed patterns of temperature change to see whether they matched the unique patterns expected to result from the combination of growing industrial emissions of heat-trapping gases like carbon dioxide, on one hand, and sulfate aerosols that cool some parts of the planet, on the other. According to this reasoning, the pattern produced by the combination of greenhouse gases and aerosols would be markedly different from that produced by any natural cause.

Dr. Santer likens this approach to that of a doctor looking for the telltale diagnostic pattern of a specific illness to explain a general rise in body temperature.

But how would one know what kind of pattern the greenhouse gas and aerosol emissions should produce? By using climate models to simulate their effect. In comparing the resulting temperature pattern with the global pattern actually observed, Dr. Santer and other researchers have found that reality corresponds markedly with the model-generated expectations. The correspondence has grown stronger over the years, as would be expected if increasing greenhouse gases were changing the climate. Statistical analysis determined that the correspondence was unlikely to have arisen by chance.

Dr. Santer and some colleagues recently published a paper in the journal *Nature* reporting a similar correspondence when vertical temperature patterns in the atmosphere are analyzed.

These pattern results, when combined with other signs associated with a human influence on climate—a rising sea level, water vapor increases over tropical oceans and more frequent bouts of extreme weather in North America, for instance—are beginning to tell "a coherent story," Dr. Santer said. This emerging portrayal was the basis for the Chapter 8 findings.

Already, he says, a number of laboratories are starting to strengthen the pattern studies by including other variables along with temperature—precipitation, for instance, and atmospheric pressure.

Researchers are also beginning to deal with the problem of improving the climate models' representation of natural variability, including, for example, changes in solar radiation. The models have been widely criticized for, among other things, failing to adequately represent natural variability. One critic, Dr. Richard S. Lindzen of the Massachusetts Institute of Technology, says the models are so flawed as to be no more reliable than a Ouija board.

"I think that's garbage," said Dr. Santer, part of whose job is to assess how good the models are. "I think models are credible tools and the only tools we have to define what sort of greenhouse signal to look for. It's clear that the ability of models to simulate important features of present-day climate has improved enormously." He says that if the models are right—still a big if—the human imprint on the climate should emerge more clearly in the next few years. All in all, he says, he expects "very rapid" progress in the search for the greenhouse fingerprint.

When might it become clear enough to be widely convincing?

"Even if New York were under six feet of water, there would be people who would still say, 'Well, this is a natural event,'" he said.

Greenhouse critics, he said, would like to "skew the focus of the science totally in the direction of the uncertainties, and they tried like heck to do it in Madrid." Having lost their case on the merits, he said, they are now trying to cast doubt on the science by attacking the scientists and the process by which the United Nations panel reached its conclusions.

Dr. Santer is still not re-engaged in the research he says he loves. But he has recovered enough to say that contrary to his first reaction, he would now take on the U.N. task again, if asked, when a new climate assessment is made in the year 2000. He would do it, he says, to counter what he believes is a systematic campaign of disinformation.

"My responsibility," he says, "is to do the best job possible explaining the science to people."

—WILLIAM K. STEVENS
August 1996

WORLD, AND DEBATE, KEEP HEATING UP

As the 20th century ended, evidence mounted that the average surface temperature of the Earth had risen by one degree Fahrenheit or more, over the last 100 years. Mainstream scientists held firmly to their prediction that it would warm substantially more in the 21st century if the world continued to burn fossil fuels at anything near the 20th century rate.

But experts had made slow progress in gauging the extent to which human activity was responsible for the warming.

Among the signs of a warming world that emerged near the turn of the millennium were these:

- Reconstructions of the Earth's climate, based on thermometer readings and "proxy" evidence of climatic change contained in tree rings, lake and ocean sediments, ancient ice and coral reefs, showed that the Northern Hemisphere was warmer in the 20th century than in any other century of the previous 1,000 years. The century's sharp jump upward ended a 900-year cooling trend. The 1990's were the warmest decade of the century.
- A variety of studies—involving diverse sources of evidence like nesting birds, satellite surveys of the landscape, the budding of trees and the shedding of leaves—revealed clearly that spring in the hemisphere was arriving earlier and autumn later.
- Other evidence indicated that in the United States, extreme summer heat and humidity most threatening to life had become more frequent over the previous half-century.

As these bits of information were emerging, an examination of ancient atmospheric gas contained in ice cores extracted from the Antarctic ice sheet showed that the present-day atmosphere contained higher levels of carbon dioxide than at any other time in the last 420,000 years—20 percent higher than in any previous warm period between ice ages and double the typical level during an ice age. Carbon dioxide, which is produced by the burning of fossil fuels like coal, oil and natural gas, is the most important of a number of waste industrial "greenhouse" gases that trap heat in the atmosphere.

But while the consensus among mainstream scientists was that the burning of fossil fuels was indeed changing the climate, experts were still unable to quantify the human contribution to the observed global warming. Benjamin Santer, who withdrew from the formal, internationally sponsored search for the greenhouse fingerprint after 1996 to return to his own research, was convinced that the human impact on climate was not trivial. Contrarians like Richard

Lindzen continued to assert that humans were not changing the climate in any significant way.

Many scientists believed it would not be possible to pin down the magnitude of the human influence until the degree of natural variation in climate was better known. In 1999, one study by scientists at the Hadley Center for Climate Prediction and Research in England found that in the early part of the 20th century, global warming could be explained either by an increase in solar radiation or a combination of stronger solar radiation and increasing greenhouse gases. But the researchers found that after the mid-1970's, when about half the century's warming took place, the warming resulted largely from the greenhouse gases.

As the search for the human imprint on climate slogged on, mainstream scientists continued to predict that the average global surface temperature would rise by about 3.5 degrees Fahrenheit (some said a little more) by 2100 if no action was taken to rein in greenhouse gases. By comparison, the temperature has risen by about 5 to 9 degrees since the depths of the last ice age 18,000 to 20,000 years ago.

The predicted warming, the experts said, could make both floods and droughts worse; cause the global sea level to rise by about 18 inches, inundating many coastal areas and island nations and dislocating tens of millions of people; shift climatic zones toward the poles by hundreds of miles; shorten winters but make heat waves more deadly; and make some northern farming regions more productive while bringing agricultural disaster to some developed countries.

Contrarians continued to insist the warming would be on the low side and would be mostly beneficial. And Richard Lindzen continued to argue that the climate system's sensitivity is such that only a minuscule warming is to be expected.

—W.K.S.

Richard S. Lindzen

It's Getting Hotter;
So What? a Skeptic Asks

Rick Friedman *for* **The New York Times**

As climate experts firm up their view that human activity is seriously altering the atmosphere, one voice stands out in clarion dissent. It is that of Dr. Richard S. Lindzen of the Massachusetts Institute of Technology, a shoemaker's son from the Bronx who has risen through the academic hierarchy as a leading expert on the physical processes of the atmosphere.

Is there truly cause to worry that emissions of waste industrial gases that trap heat, such as carbon dioxide, could disrupt the world's climate?

Dr. Lindzen does not equivocate. "We don't have any evidence that this is a serious problem," he says flatly, with precise diction, in a friendly voice that resonates strongly in his 17th-floor office overlooking Boston across the Charles River. A clutter of folders and papers, coupled with Dr. Lindzen's untrimmed black beard and horn-rimmed glasses, suggest the academic theorist he is.

His opinions attacking the formal consensus about climate change have made the 56-year-old Dr. Lindzen a bête noire to environmentalists who trumpet the dangers of global warming and a champion to political conservatives and industrial interests who minimize the threat. Admirers see him as a force for intellectual honesty in a highly politicized debate. Critics fault him for professing unwarranted sureness in a field of research rife with uncertainty. Many say he is simply wrong.

But everyone takes him with the utmost seriousness because of a reputation for brilliance that got him elected to the National Academy of Sciences at age 37. It is in this sense that his voice stands out from the relatively small group of greenhouse scientists who speak out boldly and publicly in dissent.

Last fall, a panel of scientists convened by the United Nations to advise the world's governments concluded for the first time that greenhouse gases like carbon dioxide are probably responsible, at least in part, for a changing global climate. The panel also predicted that if emissions of the gases are not reduced, the average global temperature will increase by 1.8 to 6.3 degrees Fahrenheit—with a best estimate of 3.6 degrees—by the year 2100. The predicted warming, according to the United Nations panel, would be accompanied by widespread climatic disruption.

Bunk, says Dr. Lindzen. He says the conclusions are based on computerized models of the climate system so flawed as to be meaningless. Everyone recognizes that the models are imperfect, but Dr. Lindzen goes much farther. "I do not accept the model results as evidence," he says, because trusting them "is like trusting a ouija board." Furthermore, he argues, the physics of the atmosphere permit only a minor and untroubling warming despite the buildup of carbon dioxide in the atmosphere.

His assertions have subjected him to a barrage of criticism from the modeling community and its many scientific allies. Dr. Lindzen has "sacrificed his luminosity by taking a stand that most of us feel is scientifically unsound," said a longtime acquaintance, Dr. Jerry D. Mahlman, director of the National Oceanic and Atmospheric Administration's Geophysical Fluid Dynamics Laboratory at Princeton University, which runs one of the computer models Dr. Lindzen scorns. Dr. Mahlman nevertheless describes Dr. Lindzen as "a formidable opponent."

Does Dr. Lindzen feel beaten up on?

"That's a matter of how thick one's skin is," he said over fried calamari at a Cambridge restaurant. It helps, he said, that he gets many calls and letters from silent skeptics in the scientific community who thank him for his stand. They often do not themselves speak out, he says, because it doesn't pay to be a skeptic. "Who needs to be in controversies, who needs any of this?" he said. "And in a time of budgetary restraint, climate has gotten good funding. Why spoil a good thing?"

By the same token, he says, there is a financing advantage in being on the greenhouse bandwagon. Are some people trying to maintain a sense of crisis to get research grants? "Yes," he says, "and it's unconscious and it's natural."

One who says he basically agrees with Dr. Lindzen's view is Dr. William Gray of Colorado State University, best known for his predictions of hurricane activity. "A lot of my older colleagues are very skeptical on the global warming thing," said Dr. Gray. He calls Dr. Lindzen's stand "courageous." While some of the criticisms delivered by Dr. Lindzen may have some flaws, said Dr. Gray, "across the board he's generally very good."

Another atmospheric scientist cited by Dr. Lindzen as a fellow skeptic, Dr. John M. Wallace of the University of Washington, said there are "relatively few scientists who are as skeptical of the whole thing as Dick is." Many more, said Dr. Wallace, take the question of climate change seriously but think that assertions of climate change already in progress have been exaggerated, as he does.

The object of the conflicting opinions was born in Webster, Massachusetts, in 1940, after his parents fled Hitler's Germany. His shoemaker father later moved the family to the Bronx, where, Dr. Lindzen says, "I think we were the first Jewish family in an Irish-Catholic neighborhood." There he developed a lifelong enthusiasm for amateur radio and won Regents' and National Merit scholarships at the Bronx High School of Science (class of 1956). He also acquired the middle-class native New Yorker's hardened "G" (as in Lon-GIH-land) that faintly modifies his otherwise straightforward academic accent today.

The scholarships propelled him as a student first to Rensselaer Polytechnic Institute and then to Harvard University, where he was attracted by classical physics, and then atmospheric physics. By his mid-30's he had produced landmark work in atmospheric dynamics, mainly involving "tides," or regular changes in atmospheric pressure, and the periodic shift in direction of high-level equatorial winds that affect global circulation patterns. After various academic posts, he joined the MIT faculty in 1983, where he is the Alfred P. Sloan Professor of Meteorology.

In recent years, while pursuing his main interest of atmospheric dynamics in trying to help "figure out how climate works," he has leveled a variety of criticisms at the idea of serious climatic change, some with telling effect. For instance, he points out, the computer models do not reflect the climate's natural variability very well—a key shortcoming in trying to gauge the human effect on climate, one that is readily conceded by the modelers.

But the Lindzen idea that has attracted most attention is based on a fundamental point of physics: that carbon dioxide and the other waste gases generate only a small amount of warming. Something has to amplify that warming for the larger amount of warming predicted by the United Nations panel to materialize. The main candidate, whose presumed amplifying effect is built into the computer models, is water vapor—also a heat-trapping gas, and the most powerful one since there is so much of it and it is so pervasive. The theory is that a warmer atmosphere holds more water vapor, thereby increasing the warming even more. Without this amplification, Dr. Lindzen argues, the average global temperature will rise by only about a degree Fahrenheit if atmospheric carbon dioxide is doubled.

"With all due respect to alarmism," he says, "I don't think anyone has argued that's going to be a major change in life as we know it."

The amount of warming produced by the doubling of carbon dioxide has become a standard measure of the climate system's sensitivity to change; the United Nations panel figures it at 3 to 8 degrees. But Dr. Lindzen argues that the models that produced those numbers do not properly reflect the physics of the water vapor question. While it is well known that warmer air generally holds more vapor, he says, "we don't know what determines upper-level water vapor," a factor he says is crucial, and central to the predictions of future climate change. He has postulated over the years a number of possible atmospheric mechanisms that might nullify the supposed greenhouse amplification. It was Dr. Lindzen himself who later found the first of these possible mechanisms to be trivial in its effect, but he has advanced others.

Modelers, who insist that the amplifying effect of water vapor is supported by real-world data, sharply dispute Dr. Lindzen on this point and say he has no evidence to prove his nullification hypothesis. But other atmospheric scientists say the issue is not yet resolved. "To be fair," Dr.

Lindzen says, "the answer at this stage is that we don't know" what the vapor effect is. And in fact, some important aspects of the issue are left uncertain in the United Nations panel's report.

In the meantime, he is exploring other ways to test the climate's sensitivity to change. One avenue involves volcanic eruptions. If the climate is not very sensitive, Dr. Lindzen says, the cooling effect produced by the haze from eruptions should dissipate quickly; if it is very sensitive, the effect should linger. So far, he says, it appears that the volcanoes are indicating low sensitivity—bolstering, if he is right, his overall argument about warming.

To some critics, Dr. Lindzen's confidence about the climate's low sensitivity to carbon dioxide emissions embodies more certitude than the facts allow. "I don't know what line from God he has," says Dr. Stephen Schneider of Stanford University, who cites what he sees as the overprecise estimate Dr. Lindzen gives.

Dr. Lindzen replies that he at least gives some reasons for his estimate rather than simply following a "herd instinct" that he says is very common in science. All in all, he says, "I don't think we've made the case yet" that serious climate change is in prospect, and he therefore sees no cause to worry now. To Dr. Mahlman, Dr. Lindzen is saying, "we don't know nearly as much as we need to know, therefore we shouldn't have any concern." This view, Dr. Mahlman asserts, is "deeply fallacious."

Dr. Lindzen traces his decision to begin voicing his doubts to a speaking engagement at Tufts University in the spring of 1989. After he stated his case, he recalled: "One person after another got up, saying scientists can have their doubts but we don't have any. I developed this awful feeling that here was an issue that was running away, developing a reality that transcended the science."

He says he prizes the environment, but that global warming and other issues have prompted environmental groups to go "off the deep end" and produce "a drum roll that gets rid of perspective." One consequence, he says, is that "if you are questioning the basis of global warming, you are definitely treated as someone who hates the Earth—and that's a little bit odd, because it is, after all, a scientific question."

Dr. Lindzen has always been a Democrat, too, but he says the global warming controversy has caused him to change parties. The notion that "extremely weak science" could set into motion policies with long-term implications for the economy made him "queasy" about governmental action, he said. In the academic community, he volunteers, laughing, his turn to the Republicans "must have been like coming out as a gay 25 years ago."

In 5 or 10 years, Dr. Lindzen says, direct observations of the climate's behavior might well make it clearer who is right in the greenhouse debate. Could he turn out to be wrong when all is said and done?

"I think it's unlikely," he said, "but it can happen."

—WILLIAM K. STEVENS
June 1996

Judith Lea Swain

The Double Life of Dr. Swain: Work and More Work

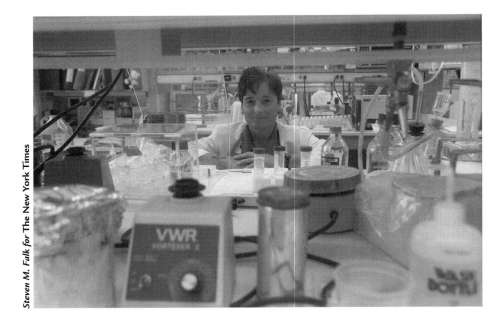

Steven M. Falk for The New York Times

D r. Judith Lea Swain, director of the division of cardiology, director of a molecular biology laboratory and professor of genetics at the University of Pennsylvania School of Medicine in Philadelphia, is determinedly, decisively, at the top of her highly competitive, male-dominated field. She planned it that way and would never accept anything less.

Dr. Swain, 45, is well known among molecular geneticists for her groundbreaking work on a surprising and significant finding called genetic imprinting. She is also the only woman to direct a cardiology department at a major medical center and, as the president-elect of the prestigious American Society for Clinical Investigation, the only woman to achieve that rank since the society was founded in 1909.

Dr. Swain is also known for having an identical twin sister, Dr. Julie A. Swain, a heart surgeon and chief of cardiovascular surgery at the University of Nevada School of Medicine in Las Vegas. Dr. Julie Swain was the first woman to head a heart surgery department at an American medical school.

Dr. Judith Swain, who elicits fervent praise from her colleagues, attributes her rise to the top to what others might describe as an almost inhuman work pace and a determination to be the best. While some might suggest that her success has come at a high cost to her personal life, she says she is driven by a simple credo: "It's not, 'Can I compete?' It's, 'Can I be the best?'" And if, as occasionally happens, she realizes that she cannot be the best in an arena, she moves on to something else.

The competitive focus on excellence began when the Swain twins were growing up in the small Southern California town of Cypress. The only children of Christine Swain, the town librarian, and John Swain, a salesman for Hancock Oil, the girls were always told that they could, and should, be best at everything they did. And, both Judith and Julie Swain say, the two sisters always competed with each other. They were star students and star athletes, and both, after watching "Ben Casey" and "Dr. Kildare," wanted to be doctors and save lives.

Judith and Julie Swain began their freshman year at the University of California at Los Angeles in 1966. The first semester was hard—their public high school had not prepared them for their rigorous courses at college—and they got C's. But they came for extra help during their professors' office hours, worked hard and by the second semester had caught up. From then on they got A's.

At the end of their first year, the twins were accepted into a special accelerated chemistry program. The program "identifies you out of 40,000 undergraduates as very special people," said Dr. Judith Swain. "We were taken care of absolutely royally." She and her sister became chemistry majors and worked in chemistry research labs during the school year and summers.

The twins separated for medical school, with Judith going to the University of California at San Diego and Julie going to Baylor College of Medicine in Houston. In San Diego, Dr. Judith Swain became a cardiologist, largely because Dr. Eugene Braunwald, who is now at Harvard Medical

School, was chief of medicine there. Her sister followed the example of Baylor's Dr. Michael E. DeBakey, a leading heart surgeon.

Dr. Judith Swain went to Duke University for her internship, and ended up staying there 17 years. She met and married Dr. Edward W. Holmes, and left three years ago when both she and her husband got positions at the University of Pennsylvania, where Dr. Holmes is chief of medicine.

Dr. Elizabeth Nabel, a cardiologist and molecular biologist at the University of Michigan, said Dr. Judith Swain is an inspiration to her. "She is probably the first very highly successful woman in academic cardiology," Dr. Nabel said. She added that to say it is difficult for women in cardiology "is an understatement."

Dr. Swain is undaunted by difficulties. "The world is very competitive, but there's always room at the top," she said.

Dr. Swain said that as she became more established at Duke, men often would ask her to be on university committees. Afraid of being perceived as a token woman, Dr. Swain would reply: "I'll tell you what. You get another woman first, and then ask me. Then I'll know you're asking me because you want me, not because you want someone with two X chromosomes."

While she was at Duke, Dr. Swain ran a basic research laboratory and worked part time as an interventional cardiologist, using high-tech approaches to treating heart disease that included balloon angioplasty and threading tiny catheters through patients' blood vessels and up to the heart. The field of interventional cardiology was rapidly advancing, "getting to be a very sophisticated business," she said. "I decided that if I was going to be in the catheterization lab, I was going to do it just as well as the four guys in the lab—or better—or I wasn't going to do it." She decided to switch specialties and went into the less technical field of coronary care.

Dr. Swain is best known for her research in molecular biology, a field that she ventured into one year while on sabbatical from her regular work. She soon stumbled on a momentous discovery.

In 1985, Dr. Swain and her husband, who both had sabbaticals, decided to learn the latest methods of molecular biology.

"My research was going well, but I wasn't sure I'd be in the forefront in five years," Dr. Swain explained. She and her husband, she said, "looked around and started at the top, writing to great people."

Dr. Swain wrote to Dr. Philip Leder, chairman of the department of genetics at Harvard Medical School and a senior investigator of the Howard Hughes Medical Institute. She followed her letter with a call to him on a Friday evening. "He answered the phone and we talked for half an hour," Dr. Swain said. Through a happy coincidence, Dr. Leder had made a new observation and he thought Dr. Swain would be the perfect person to investigate.

She spent the following year with Dr. Leder at Harvard Medical School, while her husband worked with molecular biologists nearby at the Massachusetts Institute of Technology.

Dr. Leder's project turned out to be a disappointment. He had added a cancer gene, c-myc, to mouse embryos in an attempt to understand what the gene normally does. At first, it seemed the gene was always expressed in heart muscle. But then it began to appear that the gene was only sometimes expressed there. "It looked random," Dr. Swain said, and so the investigation was terminated and she began working on another project. But she still hoped that she could learn something from the mice, and continued studying the data.

"I remember the moment," she said of her discovery. "I was sitting at my desk one evening and I realized that every time the gene was inherited from the male, it was expressed and every time it was inherited from the female, it wasn't. I called Phil and said: 'Are you sitting down? Let me tell you what I found.'"

It was the first example of genetic imprinting, a process that seems to defy the laws of inheritance. During the production of egg and sperm, a gene from the female would be chemically blocked, and that blockage would mask its effects so that the gene would be forever silent. The gene from the male would then be the only one expressed. Investigators have now found a number of rare human diseases in which imprinting hid inheritance patterns, and they speculate that more common diseases, like congenital heart disease or perhaps even some cancers might be inherited, but with imprinting.

One of Dr. Swain's current projects is to see whether it might be possible to coax new heart muscle to grow after a heart attack.

Dr. Swain explained that after about age 2, heart muscle cells do not grow. So her thought is to stimulate the cells to divide with proteins, called growth factors, that can signal cell division, or to use different proteins to convert other heart cells, the fibroblasts, into heart muscle cells.

Another research project is to study the transfer of genes to the heart and blood vessels, to try to stimulate the growth of blood vessels that can bypass clogged arteries.

With her research and her administration of the cardiology department at the medical school, Dr. Swain is essentially doing two full-time jobs.

Her day starts at 5:50 in the morning. She has laid out her own and her husband's clothes the night before, a practice that makes Dr. Holmes joke that living with her is like going to camp. By 6:15 they are in their car, eating bagels that Dr. Swain has toasted and drinking coffee that she made. Dr. Swain is driving. "I was trained as an interventional cardiologist," she said. "I'm used to doing 10 things at once."

Dr. Swain is at work at 6:30 and starts her day by having coffee with the half dozen researchers in her lab. She purposely sets aside periods for friendly chatter, as part of her effort to have a smoothly running and collegial lab. Friday nights at 5, for example, are reserved for beer and darts.

"I spend an incredible amount of time making it look like everyone is having a great time and it all is happening spontaneously," she says of her lab administration.

During the five hours she spends in the lab, Dr. Swain immerses herself in the deliberate pace of basic research. The goal, she says, is "to think and reason and not be in a rush." She takes no telephone calls and concentrates on data and what they mean.

At 11:30, she goes to her other office, at the cardiology department, where she puts on a white coat and starts a round of meetings, overseeing the work of 320 doctors, recruiting doctors in private practice to become referring physicians for the university's coronary service and recruiting others to join the department.

"I don't have a medical practice myself," Dr. Swain says. "I absolutely don't believe that a half-day-a-week doctor is one I'd want to provide my own health care."

Dr. Swain goes home at 8 P.M. on days when she has no dinner meetings. She and Dr. Holmes have dinner together, they work some more, she rides her exercise bicycle and they go to bed.

The only time when Dr. Swain and Dr. Holmes take a break from working is on Saturday afternoons when they play golf together.

Dr. Holmes said that he and Dr. Swain have "a wonderful relationship" and an intense life. When their work is going well, "there is nothing like dinner together when the two of us can share what we are doing," he said. "But the bad part is, you can't get away from it," he continued. "The highs are very, very high. The lows are very low."

Dr. Swain and Dr. Holmes have no children. "If I'm going to be a mother, I ought to be serious about it and spend the time I ought to spend," Dr. Swain said. She and Dr. Holmes have two cats "who are on autopilot." She added, "We decided that in a few years we may be settled and mature enough for a dog."

Dr. Swain does have a dream she has not fulfilled. She would like to land an airplane on an aircraft carrier. She knows how to fly a plane, just as she knows how to shoot a gun and how to parachute jump and how to excel in virtually every sport.

But, she says, "I would do anything to do a carrier landing. My goal is to do it at night, but I would take day."

—Gina Kolata
September 1994

BUSIER, AND HAPPIER, THAN EVER

Dr. Judith Swain has been defying the laws of nature. As she gets older, she is increasing the pace of her life, not slowing it down.

In December 1996, she went to Stanford University, leaving the University of Pennsylvania where she had been director of the division of cardiology for a grander position, the Arthur L. Bloomfield Professor of Medicine and chairwoman of the department of medicine at the Stanford University Medical Center. Reached at her office recently, she said her current activities included running the department of medicine, doing research in her lab, planning to start a biotech company, playing polo—a new sport for her—with the Stanford Polo Club, and organizing the department heads at the medical school to run as a relay team in a 200-mile race from Napa to Santa Cruz.

"I couldn't be happier," Dr. Swain said.

But there is that small matter of her husband, Dr. Edward Holmes, who was chief of medicine at the University of Pennsylvania when Dr. Swain worked there. How did she manage that complication?

No problem, Dr. Swain says. Dr. Holmes is now at Duke University in Durham, North Carolina, where he is dean of the school of medicine and vice chancellor for academic affairs. Dr. Swain and Dr. Holmes fly across the country to see each other on weekends—about three weekends out of four. "It's almost like dating your husband," Dr. Swain said. Whoever is host plans the social events; they have forsworn their old habit of working all weekend and instead spend their time having other kinds of fun.

Despite her delight with life at Stanford, Dr. Swain said the highlight of the last few years took place on the East Coast on June 10, 1998. That was the day she fulfilled her aspiration and landed a plane on an aircraft carrier.

As Dr. Swain tells it, her sister-in-law, who is the Episcopal bishop of Washington, set the events in motion upon hearing that Dr. Swain had this seemingly impossible dream. She contacted a parishioner, who happened to be the Secretary of the Navy. Shortly afterward, Dr. Swain got a call from the Pentagon asking if she could be in Norfolk, Virginia, the next Tuesday.

"I said, 'There couldn't possibly be anything on my schedule that would keep me from doing this,' " Dr. Swain recalled. When she arrived, she not only landed an airplane on the USS *Enterprise* but, the next day, took off from the carrier.

"It was the most unbelievable experience of my life," Dr. Swain said. Not the landing and takeoff, which she had thought would be the high points, but the opportunity to see the military in action.

"The crew on deck has an average age of 19," she explained. "It's a very demanding job. If you don't do it right you can kill people." She watched F18's take off and land from the carrier for five straight hours, from 10 P.M. to 3 A.M., with a takeoff or landing every 20 seconds. Everyone knew the assigned job and did it. "It's called the most dangerous acre of real estate in the world. But things worked flawlessly," Dr. Swain said.

"It really makes you appreciate the military, the level of organization," Dr. Swain said. "I've got a crew of 170 faculty members with 15 years each of post–high school education and they don't come together to do anything."

—G.K.

Judah Folkman

A Lonely Warrior Against Cancer

Bill Greene/Globe Staff Photo

There are few greater hazards to a scientist's career than perceiving a truth before the means exist to test it. Repeated assertion of one's insight will first invite insistent demands for proof, then skepticism, followed by silence, ridicule and often the loss of research funds.

One who has survived this grave peril, with his sense of humor intact, is Dr. Judah Folkman, director of the surgical research laboratories at Children's Hospital in Boston. Some 35 years ago, from an accidental observation made as a young researcher, he divined that tumors must somehow induce the body to build the blood vessels that feed them.

In pursuit of this insight, more than 100 laboratories and as many as 40 biotechnology companies are now engaged in identifying the chemical signals the body uses to construct and dismantle a temporary new blood supply in events like wound healing and menstruation. Cancer cells use the same system, and indeed cannot form a tumor until they have acquired the ability to switch on the formation of new blood vessels, a process called angiogenesis.

The idea of shutting down angiogenesis, by countermanding these signals, holds high promise as a cancer therapy, and the first generation of such drugs is now undergoing clinical trials. But in 1971, when Dr. Folkman began to promote the concept, the conventional view was that tumors did not need a special blood supply and that any extra vessels were just a response to inflammation.

For most of the time since, he has been a lonely voice, turned down by grant-giving committees, spurned by journal editors, lampooned by medical students. At one moment of need in 1974 he accepted a $23 million grant from the Monsanto Company. Harvard, where he has a joint appointment at the medical school, had never before accepted industry money of this magnitude and his academic colleagues roundly denounced him as selling the university's soul.

Derek Bok, then the president of Harvard, came to his support. Dr. Folkman recalled a colleague observing at the time that the university "was an intellectual wild game preserve, and it was the job of the president to see the larger animals didn't eat the smaller ones."

Escaping the Harvard carnivores, Dr. Folkman was allowed to keep the Monsanto money, which he and Dr. Bert Vallee used to isolate one of the first inducers of angiogenesis. But he continued to pay the penalty for being a surgeon trespassing on biochemists' turf.

"He's very enthusiastic and very creative but because of that he is always out on the periphery," said Dr. Douglas Hanahan, a cancer biologist at the University of California, San Francisco, who has collaborated with Dr. Folkman. "He's been very visionary about complex processes so it has been easy for biochemists to be skeptical."

Only in the last five years or so have Dr. Folkman's ideas become fully accepted and been shown, by his laboratory and others, to have a firm foundation in molecular biology. He is no longer a voice in the wilderness but principal founder of a growing field of research, one perceived in academe and the pharmaceutical industry as holding considerable promise for the treatment of cancer.

Recently Dr. Michael O'Reilly in his laboratory discovered two natural agents, known as angiostatin and endostatin, that are powerful inhibitors of angiogenesis. Many agents can make a tumor slow down for a while, until its cells develop resistance. But endostatin forces large aggressive tumors in experimental mice to shrink back down to a microscopic size at which they have no need for a private blood supply. And as Dr. Folkman and colleagues reported in November 1997, the vessel-making cells on which endostatin acts do not develop resistance.

"He saw things others didn't see," said Dr. Robert S. Kerbel, a cancer biologist at the Sunnybrook Health Science Center in Toronto, "but on top of that he had the persistence to stick with it and almost single-handedly has opened up this very substantial field of research."

Judah Folkman took an unusual route into scientific research. His father, a rabbi in Columbus, Ohio, used to visit patients in the local hospital. As a reward for good behavior the previous week he would allow Judah, 7, to accompany him. The excitement of these visits emboldened Judah one day to tell his father he wanted to be a doctor, not a rabbi, when he grew up. "So," said his father, "you can be a rabbi-like doctor."

Trained as a surgeon at the Massachusetts General Hospital, the younger Folkman soon became interested in research. While working for the Navy in 1960 on blood substitutes, he used his apparatus, a tube that held a rabbit's thyroid gland irrigated with an artificial solution instead of blood, to see if the perfused gland could support the growth of tumors. The cancer cells he seeded into the gland formed small tumors but all stayed the same size, as if once reaching a certain point they could grow no further under those conditions.

It was this observation that led Dr. Folkman to suppose that growing tumors must induce their own blood supply, probably through secreting some factor into the surrounding tissue. But it was not until 1971 that he was able to complete further experiments and publish his ideas.

Interest was not overwhelming at first. Dr. Philippe Shubik of Green College of Oxford University, who was one of the early pioneers of the field, said, "Our work wasn't ignored but it didn't excite people terribly much because we weren't saying we would cure cancer." Dr. Folkman, however, was saying just that. "What made Folkman stand alone was his saying that the key point is to target the blood vessels to stop cancer," said Dr. Robert Auerbach, an angiogenesis expert at the University of Wisconsin.

The difficulty lay in proving the concept. It took many years to learn how to make laboratory cultures of the cells that form the new hair-thin blood vessels known as capillaries. "The biochemistry that needed to be done to isolate the factors was very complex," Dr. Auerbach said. "Folkman realized he needed outstanding biochemists to do the work and he got them."

Dr. Folkman's grant application was turned down on at least one occasion by a committee of outside scientists who advise the National Cancer Institute. A former member of the institute's board said it narrowly reversed the decision after he distributed a newspaper article about Dr. Folkman's research.

"We had some grants turned down and others accepted," Dr. Folkman said. "It wasn't something I was bitter about." Still, "for 10 years there was almost nothing but criticism every time I gave a paper," he said. In 1973 he showed that tumors inserted in the clear cornea of a rabbit's eye would visibly induce blood vessels to grow toward them. "So then people said, 'OK, tumors do make new vessels, but it is a side effect,'" Dr. Folkman recalled.

A critic then showed that a crystal of uric acid placed in the cornea would do the same thing, suggesting that any kind of inflammation would induce new blood vessels. "That silenced us for a couple of years," Dr. Folkman said. "By the mid-1970's I thought the critics were right." Only later was it understood that scavenger cells called macrophages move in to mop up the uric acid, and the macrophages secrete chemicals that cause angiogenesis.

Little by little, Dr. Folkman and others managed to put together the proofs the theory required. Factors that induce angiogenesis were found and then, even more interestingly, factors that inhibit it. Making new blood vessels, rather like blood clotting, is a high-risk activity for the body and is governed by complex controls.

When new blood vessels are needed, as in menstruation and wound healing, the process is quickly cut off once the task is done. Blood vessels that get out of control invariably cause serious problems, like macular degeneration or hemangioma, a disease of newborns.

The switch that flicks angiogenesis on and off seems to be the local state of balance between inducers and inhibitors of the process. It is curious that many tumors should secrete inhibitors as well as inducers of angiogenesis. But the result, as has been known to surgeons for a century, is that a primary tumor often keeps lesser, metastatic tumors in check. When the primary tumor is removed, the metastatic tumors flourish. Presumably the balance of rival factors favors angiogenesis in the primary tumor and repression of angiogenesis elsewhere.

Though Dr. Folkman began his work by seeking inducers of angiogenesis, it is the inhibitors that are of more immediate interest for cancer therapy. A first generation of angiogenesis inhibitors, including thalidomide and a drug called TNP470, are already in clinical trials and have shown promise, thalidomide in treating brain tumors and TNP470 with hemangioma. The next generation, angiostatin, endostatin and two substances found by other researchers, promises to be much more powerful, Dr. Folkman said.

He is both impatient and fearful of the obstacles that still lie ahead. Sitting at the conference table in his suite of laboratories in a Children's Hospital research building, he talks of the frustrations faced by Howard Florey, developer of penicillin, in getting the drug into production. He has seen these obstacles firsthand in the case of Norplant, an implantable form of contraceptive based in part on a patent he held; kept out of the United States for years, the contraceptive was approved but has failed to win widespread acceptance, in part because of litigation.

He asks a colleague to bring out the apparatus in which he made his original observation of the tumors on the thyroid gland. Then he fetches a Petri dish with a chick embryo a few days old. The heart is beating and a delicate network of blood vessels has started to sprout across the yolk. A single drop of endostatin can reverse the process, forcing the capillaries to fade and the lacy network to clear.

Between the thyroid apparatus and the Petri dish, lying next to each other on the table, stretch 35 years of work. The project is close to fruition but not yet complete. "I am really worried because I have been at this a long time," Dr. Folkman said. "I have watched wonderful things in the laboratory not make it to the clinic."

—Nicholas Wade
December 1997

THE TORTUOUS PATH TOWARD A DRUG

Dr. Judah Folkman developed the field of angiogenesis research over many years and is still widely acknowledged as its leader. Yet the particular anti-cancer agents he has developed in his own laboratory have failed to progress smoothly to the clinical stage, and in 1999 the quality of Dr. Folkman's experimental work was criticized by two senior scientists.

Dr. Folkman has conceded nothing to his critics, pointing out that it often takes many years for a useful drug to be developed, and that several other laboratories have confirmed his work.

It may well be that the high claims and hopes surrounding his work make the inevitable setbacks seem more serious than they really are. Still, the setbacks are not good news.

The pharmaceutical company Bristol-Myers Squibb has abandoned efforts to produce angiostatin, the first anti-tumor agent isolated in Dr. Folkman's laboratory. Dr. Mariano Barbacid, who was the company vice president responsible for developing angiostatin and is now director of the Spanish National Cancer Institute, said that the company could not reproduce Dr. Folkman's animal tests.

Because of the public interest in Dr. Folkman's work, the National Cancer Institute in Bethesda, Maryland, decided to do its own tests of endostatin, the second of Dr. Folkman's anti-tumor agents. The NCI scientists replicated Dr. Folkman's results while working in his laboratory, but were unable to reproduce his findings independently.

The reason, in Dr. Folkman's view, is that the samples of endostatin he sent to Bethesda were destroyed by the packing material he used.

NCI officials, despite their failure to verify Dr. Folkman's work in animals, have decided to test endostatin in patients.

Meanwhile another critic, Dr. George Yancopoulos of Regeneron Pharmaceuticals, has suggested that both angiostatin and endostatin are not the body's natural anti-tumor defenses but just artifacts created by faulty experimental procedure.

The substance that Dr. Folkman calls angiostatin is a fragment of a common natural protein called plasminogen, while endostatin is a piece of collagen, another abundant protein. Dr. Folkman's theory is that the angiostatin and endostatin are cleaved from their parent proteins by special enzymes.

Dr. Yancopoulos noted that Dr. Folkman has failed to find the alleged cleaving enzymes or their genes. He also said it was common in isolation procedures to find fragments of large common proteins that appear to have some biological activity, but that the real agents are almost always rarer, more elusive proteins.

Dr. Folkman recently announced a third anti-tumor agent, this one a fragment of anti-thrombin.

Several laboratories independent of Dr. Folkman have found that angiostatin and endostatin have anti-tumor effects. But none of the laboratories has seen the dramatic, complete regression of tumors reported by Dr. Folkman.

The theory of angiogenesis still holds high promise: develop agents that attack a tumor's blood vessels, and because these blood vessels are normal cells that cannot develop resistance, the agents will be more effective than the usual

anti-cancer drugs. Dr. Folkman has played a leading role in developing this theory and in convincing others of its promise. Many candidate chemicals based on
this principle are now being tested and some will doubtless develop into useful
drugs. Whether either angiostatin or endostatin will be in this category remains
to be seen.

—N.W.

Anne Simon

The Science Adviser
to Whaaat?

Fred LeBlanc *for* The New York Times

Proposition: the creator of a cultishly adored television show is writing a new script when he hits a snag. His character, a beautiful woman scientist whose big eyes luminesce like high-beam headlights whenever anything weird happens, which is often, discovers something in an organism that instantly convinces her it is extraterrestrial.

But what can that giveaway clue be?

That, essentially, was the question Dr. Anne Simon, a distinguished plant virologist at the University of Massachusetts in Amherst, recently posed to her Biology 100 class. How about this, she suggested: DNA is made up of just four nucleotides. What if you had an additional nucleotide pair? That would show up if the scientist did DNA sequencing and all but shout, "Alien!"

Around now, a certain segment of society, people who wear T-shirts reading "The Truth Is Out There" and "Trust No One," will be nodding, having recognized a first-season episode of Fox Television's dark confection of conspiracy, fantasy and paranormalcy, *The X-Files*. Often hooked worse than any Trekkie, these people are known as X-Philes.

But Dr. Simon did not mention the DNA device simply because she saw it on television. She was the one who thought it up. File this under "Nice Work if You Can Get It": Dr. Simon serves—intermittently, in a kind of "Who you gonna call?" script-tinkering capacity—as informal science adviser to Chris Carter, the creator of *The X-Files*.

When Mr. Carter needed a little primer on chimeric organisms—which have at least two genetically distinct types of cells because of mutations or grafting—as he was writing an episode concerning a "Fluke Man," he called Dr. Simon. When he needed a scientifically plausible method for tagging people with biological markers hidden in their smallpox vaccinations, it was, again, Dr. Simon whom he asked. It was also at her urging, she said, that an episode this season called "Post-Modern Prometheus" incorporated homeotic flies, insects whose mixed-up genes do things like making legs grow out of their mouths.

Now, for those familiar with *The X-Files* and the murky travails of its two FBI agents, Fox Mulder and Dana Scully, one question is inevitable. The show features alien abductions, telepathy and other paranormal paraphernalia, massive Government conspiracies, witchy occult doings and imaginary creatures that are part human, part who-knows-what. It is high-quality television with an arty aesthetic and intelligent dialogue, but it is a hallucinogenic paranoid's dream. It sometimes draws complaints from educators that its fantasy science confuses children.

And you, one cannot but ask Dr. Simon, are a science adviser to it? Isn't that—not to go too far—a bit like being a public health adviser to a tobacco company?

Dr. Simon, a natural teacher, was not fazed.

First, she said, there is "the wince factor." She gets very annoyed at the simple mistakes made about science on television that could be corrected by asking someone like her and changing a word or two. Second, she likes being involved with the show because it is one of the very few programs on television in which scientists are generally portrayed positively and authentically, rather than as bow-tied villains.

She said: "What Chris says is that the science looking real and being real is what makes the show scary. And if the science did not look real, the show would not be as scary."

But for Dr. Simon, it seems, the gratification comes from contributing to a character close to her own heart: a female scientist—Scully, the agent played by Gillian Anderson—who may approach fantastical problems, but does so in a realistic way.

"If you have a scientist as one of the main characters," Dr. Simon said, "she has to be analyzing things correctly, and she has to be using the resources correctly, which is great. She's not the professor on *Gilligan's Island,* who's an expert on everything from nuclear physics to botany. She goes to look for help on what she doesn't know. And the people she goes to are saying things correctly; they're using the right microscopes; they're acting like scientists."

Then, of course, there is the sheer fun of it all.

In her workaday life (behind a door bearing a sign that says, "The truth is in here"), Dr. Simon is the principal investigator in a laboratory that is exploring how viruses replicate and cause disease. She is examining turnip crinkle virus, a beautifully simple virus that has provided the fodder for many papers and brought the laboratory some renown.

Dr. Simon openly loves her work, from the painfully bright, humus-scented growth chambers where the turnips infected with viruses grow to the computer where she labors over findings, analyzing all that her team, which includes her husband, Dr. Clifford Carpenter, has found.

But that is just one kind of fun. Dr. Simon, 41, is also something of a child of Hollywood. Her father, Mayo Simon, is a playwright and screen-writer; she worked her way through college as a Disneyland guide.

Through a family connection, she was acquainted with Mr. Carter before the television show was born. When Mr. Carter needed advice on how a scientist with a strange organism that could have cancer-curing properties would analyze and test it, he called Dr. Simon. She said the scientist would try to "grow it up," or culture it, in a rich broth that would look something like weak coffee and would probably test it on himself.

And so it began, a relaxed relationship that has involved almost no exchange of money and has spanned five seasons and a movie. Dr. Simon helps only Mr. Carter, not the show's other writers, but she has covered enough ground that she is planning to write a book that would be a sort of primer on molecular and cellular biology hidden inside an exploration of the science on *The X-Files*.

If the lecture Dr. Simon recently gave to her Biology 100 class is any indication, it will also be a primer on real-world scientific thinking applied to an entertaining fantasy. In an end-of-semester treat for her students (the class was suspiciously crowded, she said), she alternated clips from *The X-Files* with explanations of her modest contributions and the science they reflected.

There was, for example, the time Mr. Carter called her and asked, "How can we establish that Scully was infected with an alien virus?" That was for the episode "Redux."

"Let's say it's an alien DNA virus, and it's integrated into her genome," she told the class, describing how she had thought through the question. "How would you know you had some alien virus genome in your genome? Let's say it was every cell. How would you know that? What would you do?"

"A Southern blot," someone—an A student—called out.

"You could do a Southern blot," Dr. Simon replied, referring to a common test for checking for a match between DNA molecules. "You could take a piece of the virus's DNA, make it radioactive and single-stranded and see whether or not it hybridizes with some of your DNA. And if it does, that would tell you that that DNA is in you. And that's what Scully's going to do in this episode."

There were more tidbits. How Dr. Simon had suggested using a sea urchin larval stage, or pluteus, to depict the strange Arctic organism. And how she had proposed that if Mr. Carter wanted Government conspirators to have tagged people using smallpox vaccines, they could have used two methods.

First, they could have coated the cowpox virus with a slightly different set of proteins for each person, and Scully could catch on by finding a computer registry in 20-letter code—because 20 amino acids make up proteins, she would naturally think it was a protein sequence, Dr. Simon said. (She is particularly proud that the show used a real cowpox structural protein, astounding a colleague of hers who decided to check it out.) And second, she told Mr. Carter, a slightly different visual pattern could have been used for each vaccination, detectable with a confocal microscope.

Ultimately, Dr. Simon said, *The X-Files* adds spice to her corner of serious science, but the greatest excitement comes from the show's ability to bring a bit of serious science to the masses.

"It's a part of outreach," she said. "To think: 20 million people saw a Southern blot done!"

—CAREY GOLDBERG
January 1998

THE SPIN-OFFS ARE OUT THERE

So what if most science fiction shows and movies are much better at the fiction part than the science part? So what if most "science" depicted in *Star Wars* or *Star Trek* or *The X-Files* is not very scientific at all?

These shows can be, nonetheless, the most super-powered of vehicles for instilling real scientific curiosity, hooks to entice idle minds into exploration.

So effective are they as science sellers, in fact, that in recent years, the conversion of popular science fiction into educational science has gone into warp speed.

The granddaddy of this sci-fi cottage industry is Dr. Lawrence M. Krauss, chairman of the physics department at Case Western Reserve University and author of a 1995 best-seller, *The Physics of Star Trek*. With that book, Dr. Krauss boldly went into uncharted territory, especially considering the two tough audiences he faced: scientific colleagues who might mock him for such a frivolous venture, and Trekkies and Trekkers who would surely pillory him if he got anything wrong.

What drove him, he says, was the desire to reach a broader audience and the awareness that "teaching of any sort, especially in popular science, is really a matter of selling."

He was referring to selling ideas, but the appeal of science-fiction shows also sells books: more than 200,000, in the case of Dr. Krauss's *Star Trek* book.

In its wake, handfuls of similarly derivative books have emerged. Dr. Anne Simon came out with *The Real Science Behind the X-Files* (Simon & Schuster) in 1999. It had competition from other recent releases, from *The Science of the X-Files* by Jeanne Cavelos, a former NASA astrophysicist who also wrote *The Science of Star Wars*, to a book on the biology of *Star Trek* and even one on neuroscience called *Star Trek on the Brain: Alien Minds, Human Minds*.

Of course, much of the "science" in these shows is pure bunk, from imaginary alien organisms to strange medical phenomena. But the spin-off books take the hooey and transform it into serious questions about the real universe. Jeanne Cavelos, in a promotion for her *X-Files* book, posed such real-world questions as: "What sort of nutrition could be derived from a diet of human livers?" Dr. Simon, similarly, asks seriously whether there is, say, a scientific way to regenerate a human head. And Dr. Krauss in his physics book often focused on the riddle of why, exactly, certain phenomena depicted in *Star Trek* were impossible in the real universe.

It is also true that while the shows themselves may not teach much real science, they do seem to perform a broader, highly useful function: simply turning on people, especially young people, to the allure of science.

According to a two-year survey of more than 30,000 students in Chicago and Indiana conducted by Purdue University in 1993, *Star Trek* was by far the most influential science-promoting factor in their lives. It surpassed parents, teachers, astronauts, authors and comic books. Harry Kloor, director of corporate relations for the Purdue physics department, said at the time: "The enjoyability of the program makes the kids watch, but it's the science that makes them ask questions."

That kind of hold on the imagination is not easily shaken, even in the fact-based world of real science. Is it a coincidence, some ask knowingly, that modern-day flip-open cell phones bear such a powerful resemblance to *Star Trek* communicators, for example?

—C.G.

Eric Steven Lander

Love of Numbers Leads to Chromosome 17

Rick Friedman for The New York Times

I n the career of Dr. Eric Steven Lander, as in the new branch of biology known as genomics, the life of numbers and the numbers in life have come together.

Dr. Lander, director of the Whitehead Institute/MIT Genome Center in Cambridge, Massachusetts, is a leader in constructing a complete catalogue of the human DNA code—the genome. But he did not arrive at this position in the traditional way, for example, with a degree in biology. Only when past 30 did this curly-haired and energetic figure first crack a book in biology.

Rather, he grew up in the thrall of numbers. As a high school mathematics whiz, he was on the United States high school team that came in a close second to the Soviet team in the world mathematics Olympiad in 1974. He later trained as a pure mathematician at Princeton University. Only then did he fall in love with biology, as he spent hours talking with his brother, Arthur, a neurologist.

Biology itself has also been undergoing change in recent years. The old style of academic biology is now admitting a brash new branch of inquiry, one that is information-heavy, computer-driven and closely allied to business. And for Dr. Lander, that has been perfect. When he emerged from his personal transformation, there he was, at the leading edge of molecular biology.

He established his credentials in biology by tackling subjects that could only be approached by someone with a strong background in mathematics, like how to analyze statistically whether a disease may be caused by one or many genes, and how to ferret out the different contributing genes.

In August 1996, a team led by Dr. Lander found a gene that contributes to Type 2 diabetes, a disease caused by many genes, each with many variants. Dr. Lander's strategy began with the calculation that elusive genes are easier to identify in isolated populations, where people are descended from only a few founders and have not accumulated the many genetic variations of more cosmopolitan groups. He searched for the diabetes gene among a group of people in the Bothnia region of western Finland, where few outsiders have migrated in the last 1,000 years.

When biologists began to consider the task of making a complete catalogue of the entire three billion letters in the human body's DNA code, Dr. Lander's work made him a natural candidate to lead one of the several teams of DNA sequencers.

Dr. J. Craig Venter, head of the Institute for Genetics Research, a private concern in Rockville, Maryland, a competitor of Dr. Lander in the race to sequence genomes, said: "In sequencing whole genomes the breakthrough has been mathematics, applied math and new algorithms. These are the kind of things Eric is good at."

At the Whitehead Institute/MIT Genome Center, Dr. Lander's group has produced the first genetic maps of the human and mouse genomes, a necessary step toward working out the complete DNA sequence. His laboratory is one of several that are financed by the National Center for Human

Genome Research in Bethesda, Maryland. The consortium of laboratories had planned to complete the full DNA sequence of the human genome by the year 2005 at a cost of $3 billion, but is already two years ahead of schedule and below budget. The project has already identified many genes of medical interest and prompted investments by several companies.

Dr. Lander, 39, was born and raised in Brooklyn in a family of lawyers. As a student at Stuyvesant High School in Manhattan, he was sent one summer to participate in an elite mathematics program, where the students decided that 17 was the most interesting of all numbers. They formed a 17 club and made up a T-shirt emblazoned with amazing facts about the number 17. Dr. Lander can still quote examples: "Many multi-sided figures are stable when set down any one of their sides, for example, a pyramid. But did you know that a 17-sided figure is the only one that is stable on one side only?"

Recently, the number 17 has sneaked back into his life. The Whitehead genome center has chosen human chromosome No. 17 as the one it will sequence as its contribution to the Human Genome Project. "Someone suggested I had picked chromosome 17 because of my fascination with that number," Dr. Lander said. "That's not really true, but I am thinking of taking the old T-shirt out of the closet. I still have it."

As Dr. Lander followed his instincts, his career took some sharp turns, from pure mathematics at Princeton and Oxford, to managerial economics at the Harvard Business School. Then, while teaching mathematically oriented business classes by day, at night he crossed the Charles River to hang out in biology laboratories.

He had begun to see that beneath the surface of the two very different disciplines of mathematics and biology there lay some links of possible importance. Biology, however chaotic it might appear, had regions that he felt would yield to the firepower of mathematical methods. His first few papers exploring mathematical approaches to biology were sufficiently remarkable that he won a MacArthur Fellowship, the so-called "genius" award. "That grant was crucial for me," he said. "I was struggling to establish myself at the interface of math and molecular biology. Why should anyone take me seriously? The MacArthur gave me that essential credibility."

The $250,000 grant helped finance travel to the far-flung and isolated human populations where he knew gene-hunting would be easier.

Dr. Lander soon started to make an impact in molecular biology, creating the mathematical tools to tease out a major gene in asthma, and a "modifier" gene that can suppress colon cancer. But eventually he tired of hunting down genes in the genetic jungle, one by one. "That time is over," he said. He is now laying plans for the next era in biology, in which he foresees that the entire set of human genes and their functions will be available on one CD-ROM disk, so there will be no more Stanley-and-Livingston searching.

"Now, suddenly, biology is finite," he said.

"The genome project is wholly analogous to the creation of the periodic table in chemistry," Dr. Lander said. Just as Mendeleev's arrangement of the chemical elements in the periodic table made coherent a previously unrelated mass of data, so Dr. Lander believes that the tens of thousands of genes in present-day organisms will all turn out to be made from combinations of a much smaller number of simpler genetic modules or elements, the primordial genes, so to speak. He theorizes that these modules helped carry on life in the most primitive cells living on the planet three billion years ago. The basic functions of life carried out by the first genes must all have been formed very early in evolution. Dr. Lander surmises. Most present-day genes are variations on these few original themes, he said.

"The point is that the 100,000 human genes shouldn't be thought of as 100,000 completely different genes," Dr. Lander said. "They should be thought of as maybe a couple hundred families that carry on essentially all of life."

Making such a periodic table for families of genes will define a new direction for biology, in Dr. Lander's view. The completed table would mark the end of structural genomics, the analysis of the structure of genes. "When you get the last base of the genome, driven in like the golden spike in the transcontinental railroad, we'll maybe have a big ceremony," he said. "But when it's done, it's done."

Then comes what Dr. Lander calls functional genomics, or making practical use of the table. For example, Dr. Lander says, biologists may learn to read human DNA so effectively that laboratories will quickly be able to tell patients all the important variations they have in their entire gene sets. Further, it should be possible to tell which of those genes are turned off or on at a given moment, thus getting a picture of whether the cells of the body are up to snuff.

"So here's the manifesto for the era of functional genomics," Dr. Lander said. "One. At the DNA level we want the ability to re-sequence an entire genome—anybody's genome—in a regular medical setting, to find all the variations. Because you and I differ in one-tenth of 1 percent of our bases, and that accounts for our differences. Most genes will have two, three or four major variants. If you have 100,000 genes, that means there will only be about 300,000 major variants. It's a finite number. We can then take that list, and then correlate all the different variations with health outcomes. You could take the Framingham Heart Study and find the rate of each disease associated with each of the 300,000 variants of genes."

That would allow each person to get a full list of what diseases they are most at risk for, based on their inheritance. With a mix of hope and skepticism, Dr. Lander said: "In principle, that would allow us to have personalized health care and personal health care strategies. In practice, of course, whether we do that will depend on what we as a society want to pay for, and how much we can protect our privacy, and so on."

"Two," he said, holding up fingers to signal the next item on his manifesto. "We want to be able to monitor gene expression." Finding out which of an individual's genes are active at any time would help indicate a body's response to drugs, dieting, exercise and other factors.

"All this is not so crazy as it sounds," Dr. Lander said. "Less crazy, in fact, than the genome project itself. There are already genetic 'chips' that can make these things possible."

He was referring to one of his favorite new technologies, which has put human genes on microchips. Genes in a blood sample can be matched against the standard ones on the chip to see if there are any important abnormalities. So far, one company making "gene chips," Affymetrix Inc. of Santa Clara, California, has succeeded in putting all the genes of HIV, the virus that causes AIDS, on a chip for such comparison. The company has plans to put 30 to 40 human genes on one chip, and "in principle at least," said Robert Lipschutz of Affymetrix, "we should be able to put all human genes on a chip."

Dr. Lander has a piece of that company, as well as a major financial interest in Millennium, a company that intends to make use of the data from the genome project to design diagnostics and treatments for disease.

If there is a danger sighted ahead in the "new biology," some critics suggest, it is that businesses may be too close to science, and may even sometimes be in the driver's seat. Scientific judgments may too often yield under pressure from business needs.

Dr. Lander, an avid businessman, takes these problems more seriously than most people in science, said Dr. Francis Collins, director of the Federal genome project. Dr. Collins credits Dr. Lander with leading the way to help solve at least one of the problems—that of hoarding data to gain business advantages.

The Whitehead genome center, at Dr. Lander's direction, puts out on the Internet all the data it produces on DNA markers and sequences, which are freely available to anyone who wants to copy the material.

At first the MIT laboratory's data were posted every few months. Now they are posted monthly, and soon they will be disseminated almost daily, Dr. Lander said. "This work is paid for with public money and it's got to be made public as fast as we can," he said. "That means breaking with tradition and getting it out there long before it can be published in scientific journals."

The effect, he says, is highly stimulating for biologists. "We get 50,000 to 100,000 hits on our database per week. People need this data."

The Federal genome project office has begun to follow his lead, and those receiving grants must now make their data available at least every six months.

The task over the next few years for those leading molecular biology will be to get biologists away from their traditional tools—pipettes, gels and flasks—and into analyzing gene function with computers.

"In the next one to three years, we have to figure out how to get humans out of the loop," he said. "Then we can really get to work thinking about biology and what's going on in life."

—PHILIP J. HILTS
September 1996

TRUTH-TELLING THROUGH STATISTICS

A wellspring of data has begun to gush from the human genome—so much data, in fact, that not long ago some scientists were predicting that no one would know what to do with it all for decades to come.

But even before the complete reading of the human genome has been finished, a striking series of experiments has shown that the data can be used in powerful ways, immediately.

For years, doctors have been able to distinguish between two similar forms of leukemia only by conducting four tests in separate, highly specialized laboratories, plus studying the similar-looking cells under a microscope. Even with multiple checks, errors were common. The correct diagnosis is crucial, because the most successful drug treatments in each case are quite different.

Now, scientists at the Whitehead-MIT Center for Genome Research and the Dana-Farber Cancer Institute have shown that the difficult diagnosis can be made simply and efficiently, through a marriage of DNA technology and statistics. The experiment suggests that in some cases it will not be necessary to rely on medical judgments, which are ultimately subjective, to piece together patterns about how cancers behave.

Even more important, it appears that the new technique can be used to test for many kinds of cancer and, outpacing decades of human effort, to identify new cancer types and subtypes. The payoff should be cancer treatments that work better because they are focused more accurately.

The behavior of a cell is governed by which genes, amid the 100,000 genes residing on a cell's DNA strands, are turned on, and to what degree. The array of genes that are lighted up or turned off is quite different from cell to cell: tumor cells are different than healthy ones, and among tumors of the same cancer—say leukemia—different subtypes behave very differently as well (and thus require very different therapy).

The researchers, led by Dr. Eric Lander of Whitehead and Dr. Todd Golub of Dana-Farber, began with 38 patients who had been carefully diagnosed with either acute myeloid leukemia or acute lymphoblastic leukemia. The doctors selected 6,800 genes—all the genes easily obtainable—from a normal cell and laid them out in arrays on a microchip. They then took extracted genetic material from a patient's tumor cells and injected it onto the chip.

The chip was designed to sketch a genetic "profile" of the tumor cells, showing which genes were turned on, and to what extent those genes were active. In analyzing 36 of the tumor samples, the chip yielded two distinctive profiles, which matched the diagnoses of the two leukemia types. In two cases, the result was mixed. (As it turned out later, there was good reason for the two to come out "mixed"—there were problems with the initial diagnosis of leukemia.)

In the second step, the researchers took tumor samples from 34 different patients, whose diagnosis was unknown to the scientists when the lab test was done. From a mass of numbers identifying which genes were active, the chip-and-statistics method accurately "discovered" that there were, indeed, two main categories of disease. The method put 29 of the patients into the two categories, which was a perfect match for the leukemia diagnoses done by the more laborious traditional methods. Results in 5 of the 34 cases were left uncertain.

The experiments, which were reported in the journal *Science* in October 1999, also yielded an unexpected discovery. Some genes are so active and so closely linked to the myeloid or lymphoblastic type of leukemia that they may be useful as markers for each cancer. The active genes may also be studied for their role in the process of making a cell cancerous.

As more genes are identified through the sequencing of the human genome, scientists expect to use this new kind of genetic analysis in other cancers. Diagnosis will be quicker, and researchers will be able to produce an ever larger gallery of "wanted posters" for various types and subtypes of cancers.

—P.J.H.

Kathy Schick and Nicholas Toth

Recreating Stone Tools to Learn Makers' Ways

Mark Simons

In the summer of 1993, Kathy Schick and Nicholas Toth drove out to Oregon, bought 2,000 pounds of obsidian, quartzite and basalt from three quarries, hauled it back to their home here in a rental truck and dumped it in their backyard. For Dr. Toth and Dr. Schick, a husband-and-wife team of archaeologists at Indiana University at Bloomington, the load of rocks is a vehicle in interpreting human evolution.

Over the next few years they, their students, volunteers and a pygmy chimpanzee named Kanzi, well known in the anthropology community for his participation in language experiments, will flake the stones into tools, just as humans' hominid ancestors did beginning 2.5 million years ago.

The lithic experiments are part of their project to investigate human origins and evolution through technology.

"If anything defines the human condition, it's tools and technology," said Dr. Toth, who, with Dr. Schick, founded the university's Center for Research into the Anthropological Foundations of Technology in 1986, when they were junior members of the anthropology faculty. As hominids started using tools and becoming carnivores, they began to "produce simulated biological organs—slashing, crushing organs," Dr. Toth said.

"You see a reduction in the size of the teeth and jaws because we're replacing biology through technology," he continued.

In recent years their varied research has included investigating the toolmaking ability of a pygmy chimpanzee, again using Kanzi, and reinterpreting the uses of the earliest recognized stone implements, those first identified at Olduvai Gorge in Tanzania and known as Oldowan tools. Since 1989, they have participated in several excavations in China as part of the first American archaeological team permitted to work there in more than half a century. Their book, *Making Silent Stones Speak: Human Evolution and the Dawn of Technology* (Simon & Schuster, 1993), got favorable reviews.

Dr. Toth and Dr. Schick began collaborating in 1976, when they met on a dig in their home state, Ohio. They married a year later, when they were graduate students at the University of California at Berkeley, and spent half of each of the next three years living in a two-person tent at an early hominid site in the badlands of Koobi Fora in Kenya.

"You couldn't slam the door after a fight," said Dr. Schick, who is 45.

In fact, they enjoy working together and believe their ability to evaluate each other's work frankly and rigorously improves their research.

"I don't think we can be so brutal to our colleagues," Dr. Toth said.

The most important aspect of their research, and one they have carried further than other researchers, is experimental archaeology. As a graduate student, Dr. Toth, 42, realized that to find out who was making the stone tools, he needed to understand how they were made. Building on the work of a few other lithic tool makers, he began flaking stones and testing out the products. Dr. Schick soon joined in these efforts, which included butchering various animals, including a few elephants. (All had died of natural causes.)

From the experiments, Dr. Toth developed an important hypothesis about Oldowan tools, the simple flaked stone implements that first appeared 2.5 million years ago and were named by Dr. Mary Leakey and Dr. Louis Leakey, the archaeologists.

"The easy but now erroneous inference was that because of their different shapes, the cores were the tools," Dr. Schick said of these implements.

But in their experimental butcheries, Dr. Toth and Dr. Schick found that the small flakes struck from a core were much more effective in cutting than the core itself. With an obsidian flake they could slice through the one-inch-thick hide of an elephant. When the blade dulled, after about five minutes of work, they simply hammered on the core with a large stone to get another flake.

"When people think of stone they think 'primitive, not functional,' but you can't get anything sharper than a flake," Dr. Toth said.

Far from being waste products, known as debitage, the flakes were at least as useful to early hominids as the chunky cores, they concluded.

Their experiments also shed light on the brains of the early hominids—probably *Homo habilis*—who used these tools. Dr. Toth noticed that flakes struck side by side from a core exhibited different features depending on whether the tool maker held the stone hammer in his or her left or right hand. Analyzing the flakes he had produced, Dr. Toth, who is right-handed, found that of those that showed such differences, 56 percent of them had right-handed features and 44 percent of them left-handed features. He established this ratio as being consistent with the pattern produced by a right-handed person. An analysis of flakes found at Koobi Fora showed a ratio of 57 to 43. From this Dr. Toth inferred that the hominids who made the tools had been preferentially right-handed. Modern-day humans are 90 percent right-handed and are the only known creatures to show such a pronounced preference for one hand. Dr. Toth's work suggests that for whatever reason, the brains of these tool makers were already profoundly lateralized.

Some scientists had suggested that the earliest hominids were probably similar intellectually to modern-day chimpanzees, and this idea prompted another project.

"If you could teach a chimp to make stone tools," Dr. Schick said, "how would they compare to the earliest stone tool record?" In 1990, Dr. Toth and Dr. Schick, working with psychologists at the Language Research Laboratory, operated by Georgia State University and Emory University in Decatur, Georgia, devised an experiment to see whether Kanzi, the pygmy chimpanzee, could make a stone tool. After demonstrating how to produce a flake by striking a stone core with a rounded hammer stone, Dr. Toth used the flake to cut a finger-thick rope that closed a clear door to a box containing fruit. Once the rope was cut, the door could be opened and the fruit removed.

"What we did was give him motivation," Dr. Schick said. Within a month Kanzi succeeded in producing a small flake, which he used to cut through the rope. Weeks later he came up with his own technique for making flakes. He began hurling the stones against a tile floor. Usually a good flake would fly off after two or three throws and he would happily pick away at the rope to get the fruit.

This surprising and amusing development temporarily frustrated the archaeologists, who have since tried to devise ways of getting Kanzi to knap instead. The flakes he has produced by knapping, while functional, are smaller than those made by early hominids. While Kanzi has finally learned to hit the tool near the edge, he does not seem to consider the edges' angles. Dr. Toth has found that early hominids struck only those edges with angles of less than 90 degrees.

"The Early Stone Age record is still pretty sophisticated compared to what Kanzi has been able to do," said Dr. Schick. "We think these early stone tools do show a cognitive advance over our nearest living relatives."

The results with Kanzi hint that there may be an earlier generation of hominid-made stone tools that has not yet been recognized by archaeologists. Dr. Schick said future research will look at the pattern of flakes produced by Kanzi's tool-making efforts. Called "site formation analysis," the approach allows archaeologists to hypothesize what activity might have occurred at a certain site by studying how fresh sites decay and become buried.

Like their mentors, Dr. J. Desmond Clark, Dr. F. Clark Howell and the late Dr. Glynn Isaac of the University of California at Berkeley, Dr. Schick and Dr. Toth are committed to their students. It is noteworthy that four of their six graduate students are from countries with important early hominid sites. In 1992 one of them, Tadewos Assebework of Ethiopia, working with Tim D. White, a paleontologist at the University of California at Berkeley, uncovered several teeth of what was announced in September to be the oldest known hominid, *Australopithecus ramidus*.

"Particularly because there's so much of an interest in these countries, they're making a substantial contribution that goes beyond the normal training, and that's going to have an effect even generations from now," said a friend, Dr. Lawrence Keeley, an anthropologist at the University of Illinois in Chicago.

In one recent class, a laboratory on stone-tool technology, Dr. Toth was reviewing tool types for an exam when a student raised a question about a dihedral burin, a common tool used for engraving 40,000 years ago.

Dr. Toth strode into the adjacent stone-tool workshop and returned with a piece of flint and a hammer stone. The students were familiar with this routine but nevertheless watched in awe as he handily scraped the edge of the flint, preparing a platform on which to strike, and then, with a firm whack, struck off a small sliver of stone with sharp points.

"The coolest thing is that when somebody doesn't understand something, he just goes and makes it," said Kim Lewis, a junior who said she became an anthropology major after taking Dr. Schick's introductory anthropology class. "I take anything they teach."

—BRENDA FOWLER
December 1994

HIGH-TECH ENTERS THE STONE AGE

In their continued efforts to understand the development and evolution of the simplest human tools, Kathy Schick and Nicholas Toth have begun subjecting themselves to high-tech diagnostic tools. In one experiment designed to identify the major areas of the brain used in making stone tools, Dr. Toth was hooked up to a positron emission tomography (PET) scanner, which maps the area of brain activity. He then performed three activities: he rested, in order to provide a baseline for activity; then he thought about making stone tools; and then he took a hammer and a cobble in his hands, and began knapping.

"Some of the areas of the brain that light up are the areas of the brain that have expanded with evolution," Dr. Toth said. They include part of the neocortex, which regulates motor coordination and orientation in space. Such experiments may also show whether there is a correlation between tool making and other cognitive skills, like language.

Working with Dr. Mary Marzke, a physical anthropologist at Arizona State University who studies the evolution of the hand, and Dr. Ron Linscheid, a hand surgeon at the Mayo Clinic in Rochester, Minnesota, both Dr. Toth and Dr. Schick have volunteered their bodies in experiments designed to evaluate which hand and arm muscles are being used in the making of a tool. Using a needle several inches long, a hand surgeon deposited wires into their forearms and hands.

"It's theoretically minimally invasive, but by the time you have all these needles and wires inserted in you you feel like a human pin cushion," Dr. Schick said.

Dr. Marzke then measured the current in a given muscle to see whether it was important in the activity.

"It had often been said over the years that the precision grip between the thumb and the tip of the forefinger was critically important, but it was not called upon that heavily," Dr. Schick said.

Instead, Dr. Marzke found that the grip between the thumb and the first knuckle on the index finger and between the thumb and the rest of the palm were more important in tool-making. Using that information, she is now looking for indications of tool use in the hand bones of early hominids.

As for Kanzi, the bonobo that Dr. Schick and Dr. Toth began teaching to make stone tools about 10 years ago, there has been progress. His knapping skills have improved in the last several years, and he now will make a few flakes upon the request of the people who work with him closely. The archaeologists have also begun training Kanzi's sister, Pan Banisha, to knap flakes, and her son as well. But Dr. Schick said that the hominids who made the earliest stone tools, who were separated by our ape ancestors by only two and a half million years or so, were still better at handling the stone than Kanzi and his sister.

"Individually some of his artifacts could compare to some of the earliest cores or flakes, but as a population his still have less refined manipulation of the core," Dr. Schick said.

The gap between the products of the bonobos and the earliest stone tools suggests that archaeologists do not understand the earliest chapters of stone tool use. "Even the earliest stone tools are surprisingly sophisticated, so we'll have to see whether there are stages of development leading up to that that may mirror some of the stages of tool making that Kanzi is going through," she said.

Though they still have nearly 20 years before retirement, Dr. Toth and Dr. Schick are already thinking about their legacy. Aided by the happy economy, the couple has devoted much of the last few years to establishing an endowment for the archaeologists on their heels as well as a monograph series on human evolutionary studies at the Indiana University Press.

"The money we raise goes directly into helping our students go into the field, and we have some great students," Dr. Toth said. Among them are students from China, Algeria and Ethiopia, countries with important early human sites.

—B.F.

James W. Cronin

Looking for a Few Good Particles From Outer Space

Steve Kagan for The New York Times

At 67, with a great career as a particle physicist and teacher already behind him, Dr. James W. Cronin of the University of Chicago has embarked on a new one as a kind of traveling salesman. He is campaigning in a score of countries for help in building a pair of gigantic cosmic-ray telescopes expected to reveal some tantalizing secrets of the universe.

The son of a professor of Greek and Latin, Dr. Cronin has devoted his career to particle experiments at the Fermi National Accelerator Laboratory in Illinois and the Brookhaven National Laboratory in Long Island, where in 1963 he and Dr. Val L. Fitch, both professors at Princeton University at the time, discovered a type of particle decay that may explain why the universe survived its violent birth instead of annihilating itself. Dr. Cronin and Dr. Fitch were awarded the 1980 Nobel Prize in physics for the achievement.

Dr. Cronin has spent most of his life searching for subtle insights hidden in thickets of data, and he still loves data and the experiments that produce it.

But in recent years he has also become a spokesman for cosmic-ray research, traveling endlessly and bringing his charm and gift of persuasion to bear on government agencies and institutions in 19 nations. For what he is trying to do he needs money, equipment and scientific talent. No longer a full-time laboratory denizen, he spends as much time on airliners as in the comfortable apartment overlooking Lake Michigan that he shares with his wife, Annette, at the University of Chicago.

"Sometimes I have successful fund-raising trips and sometimes I don't," he remarked with a wry smile. "I've become the Willy Loman of cosmic-ray physics."

Dr. Cronin is selling a kind of particle physics that helps explain what makes the universe tick.

Particle physicists learn from nature by accelerating specks of matter—protons and antiprotons, for example—and forcing them to collide with fixed targets or with each other. These collisions produce flashes of energy that convert themselves into new particles of matter, many of them bizarre forms that existed in nature for only an instant after the Big Bang of creation.

But Dr. Cronin has turned away from laboratory accelerators to focus his attention on the sky, which furnishes a constant rain of cosmic-ray particles, some with almost inconceivably great energies. The half-dozen most energetic cosmic-ray particles detected so far packed a punch of more than 10^{20} (100 quadrillion) electron volts each, about 10 million times more energy than that of a particle boosted by the world's most powerful accelerator. This energy is comparable with that of a hard-hit tennis ball and no one is sure how such vast energy is generated.

The mystery of ultra-high-energy cosmic rays began piquing Dr. Cronin's curiosity a decade ago, and it has spurred him to organize an experiment on a gigantic scale designed to operate on a very tight budget—far more frugal than the budgets of most laboratory accelerators. It is called the "Auger Project" to honor the late French cosmic-ray physicist Pierre Auger (pronounced oh-ZHAY). When (and if) it is completed, it will consist of 1,600 detectors scattered across a region of Argentina the size of

Rhode Island, and of an equal number of detectors in a tract at the Dugway Proving Grounds 75 miles southwest of Salt Lake City.

Each of the detectors being built for the experiment is a light-tight metal tank containing 12 tons of water and three photomultiplier tubes. When a high-energy particle hits the water, it will leave a streak of blue light along its track—a kind of optical shock wave. The photo tubes will record the direction from which the streak originated, and with luck, the experimenters may identify some of the celestial objects from which high-energy cosmic rays are arriving.

The Argentine site near San Rafael in Mendoza Province, and the Utah site, are both sparsely populated.

The detectors will be installed in hexagonal patterns, each detector spaced from its nearest neighbors by 1 mile, and each array covering an area of 1,160 square miles. Dispersed among these detectors will be some other instruments known as "fluorescence detectors," or "fly's eyes," so named because of the multiple phototubes they point toward the sky. During clear, moonless nights, these detectors can see the faint flashes of fluorescence in the sky caused by showers of high-energy particles, and will measure the energies of these flashes.

Arrays covering huge areas on the ground are essential to the Auger experiment because it will look for the rarest of cosmic rays.

Physicists have determined that about once a century a single cosmic-ray particle with an energy of 10^{20} electron volts will hit any given one-square-kilometer (0.39 square mile) patch of Earth.

By using an area much greater than one square kilometer, the odds of snagging the shower produced by a very high energy particle are improved. When both the Argentina and Utah arrays are working, Dr. Cronin calculates, they may detect and measure about 30 ultra-high-energy cosmic-ray particles a year. Small though that harvest will be, it may help to resolve a scientific mystery.

In 1966, an American, Dr. Kenneth Greisen of Cornell University, and two Soviet physicists, Dr. G. T. Zatsepin and Dr. V. A. Kuzmin, independently reached the conclusion that no cosmic-ray particles reaching the Earth could exceed 5×10^{19} electron volts. Their reasoning was that cosmic-ray particles above this energy speeding through intergalactic space would interact with photons of microwave radiation left over from the Big Bang, and so be destroyed and converted into secondary particles of lower energy.

Sparse evidence collected since then has suggested that there might indeed be some kind of energy cutoff. But a paper published on August 10 in the journal *Physical Review Letters* by a team of Japanese physicists reveals new evidence that the hypothesized cutoff may not exist, and that there may not be any limit to the amount of energy a cosmic-ray particle might have.

The Japanese group, led by the Institute for Cosmic Ray Research of the University of Tokyo, reached its conclusions from data gathered by the Akeno Giant Air Shower Array from 1990 through 1997. This Japanese array, which covers an area of 100 square kilometers (38.6 square miles), saw a number of cosmic-ray showers with energies of magnitudes supposedly ruled out by the cutoff.

So is there a continuous range of cosmic-ray particle energies, or is there a gap in the range? Is there an upper limit to cosmic-ray energies? Where do these celestial bullets come from, and how are they accelerated?

The Auger detector arrays in Argentina and Utah, which will be the largest ever built, may solve some of these questions.

But such arrays are expensive. Dr. Cronin and his collaborators expect the Argentine and Utah detectors and apparatus to cost about $100 million. So far, there is a big deficit in financing for the Argentine array, and little if any financing for Utah.

Holding an international coalition together has also proved tricky, Dr. Cronin said; if one nation expresses misgivings about its financial contribution, others begin to worry. That means that Dr. Cronin must constantly be on the move to plug holes in the organizational dike. Currently, for example, he is conferring with officials in Argentina to strengthen their resolve.

"It helps being a Nobel laureate, because potential supporters take you more seriously," he said. "But it's tough going, even so."

Having technically retired two years ago, Dr. Cronin is now a professor emeritus at the University of Chicago. But this fall he will begin teaching part time at the University of Utah.

"I have an ulterior motive in going to Utah, besides wanting to teach there," he said. "I hope to beef up support for the Utah part of the Auger project, which has not yet been funded."

The discovery that brought fame to Dr. Cronin and the professional status helping him to attract financing had to do with the juxtaposition of matter and antimatter created in the laboratory.

The conventional wisdom of particle physicists prior to 1956 assumed that matter and antimatter were identical in every respect, except that a particle of antimatter is a mirror reversal of all the dimensions of its matter counterpart. But in 1956, faith in this assumption was shaken by two physicists at the University of Chicago, Dr. Chen Ning Yang and Dr. Tsung Dao Lee, who found evidence that this might not be strictly true. A year later they were awarded a Nobel Prize.

In 1963, Dr. Cronin and Dr. Fitch performed a very delicate experiment that conclusively demonstrated an even more striking violation of these symmetry principles, and provided an explanation for the seemingly improbable existence of our universe, which is made entirely of matter, without any antimatter.

A particle of matter that encounters an antimatter particle causes the annihilation of both particles and the creation of gamma rays that race away at the speed of light.

It was generally believed that the Big Bang created matter and antimatter in precisely equal amounts. But if that were so, theorists wondered, why had not the primordial matter and antimatter canceled each other out completely, leaving only an invisible glow of gamma rays?

The answer, Dr. Cronin and Dr. Fitch learned from their particle-beam experiment at Brookhaven, was that nature is even less even-handed than had been supposed. In technical terms, the two physicists discovered that the "charge conjugation" and "parity" observed in the decay of unstable particles called neutral K-2 mesons are slightly violated by the "weak" nuclear force. To a physicist, CP-violation, as this effect is commonly called, is disturbing; it is a little like finding a flaw in the symmetries of Newton's laws of motion. The discovery entailed hard, uncomfortable work for the two physicists and their graduate students, Dr. Cronin said. During their research in 1963 they labored in an instrument-crammed nook next to the big Alternating Gradient Synchrotron accelerator at Brookhaven National Laboratory in Long Island. Their research quarters were confining, dirty and stifling hot; "We called it Inner Mongolia," Dr. Cronin recalled, "but we loved it, and I loved the data it generated. In fact, I love data in general."

The 1963 discovery of CP-violation at the microscopic particle level offered a possible explanation of why the universe was not annihilated soon after its birth. The amounts of matter and antimatter created by the Big Bang must have been nearly equal, but not precisely so. For every billion particles of antimatter created in the genesis explosion, the experiment by Dr. Cronin and Dr. Fitch demonstrated, there could have been a billion and one particles of matter. One surviving atom of matter for every billion particles annihilated was enough to create the universe.

Big accelerators have become so expensive to build that energies much greater than those already operating may not be attainable. In 1992, construction of the Superconducting Supercollider (SSC) was halted by a Congress unwilling to support the $8 billion project, and physicists have been forced to set their sights much lower.

"We can't match the experiments that might have been possible with the SSC," Dr. Cronin said, "but there's enormous satisfaction in studying cosmic-ray particles. The money and scientific teams are much smaller than those needed for accelerator experiments, so it's a friendlier environment. It gives scientists from many developing countries a chance to do research in high-energy physics. And it's nice doing experiments out of doors."

—MALCOLM W. BROWNE
August 1998

BIG SCIENCE FORGES AHEAD

Jim Cronin has more time for science these days. A decade of playing the role of traveling salesman, hawking his $100 million cosmic-ray project, has paid off with promises of funding from institutions in the United States as well as three nations in Latin America and five in Europe.

Completion of the Pierre Auger project, which will consist of huge cosmic-ray detectors in Argentina and Utah, now seems assured, and Dr. Cronin can focus anew on the experiment's scientific goals.

He has also devoted some time to the education of non-scientists seeking to learn something about astrophysics.

"I've been giving a series of lectures at the Collège de France in Paris on the history of particle physics," he said. "It's a really nice position at an institution that was founded in 1529 by King Francis I to bridge the intellectual gap between academics of his day and the general public. We still have to try to bridge that gap."

Meanwhile, Dr. Cronin shuttles between his apartment in Chicago, his teaching position in Paris and the site in Argentina near Malargue where the first 40 cosmic-ray detectors are being erected. Each one is a metal tank containing 12 tons of water and light sensors to detect the flashes resulting from the impacts of cosmic rays.

"I have to admit that these long airline flights in steerage-class seats are tiring me," he said.

"The first 40 detectors being erected in Argentina will serve as a prototype for the full-scale array," he said, "and that will allow us to check the system out before completing it." The full Argentine array is expected to start operation in 2003.

One thing has become clear in the last several years, he said: against the expectations of theorists, ultra-high-energy cosmic rays exist, and they hit the Earth fairly regularly.

"Japanese detectors have seen at least seven cosmic-ray impacts with energies greater than 10^{20} electron volts," he said. "A detector in Utah has registered as many as seven such events."

Theorists had believed that cosmic-ray particles with such enormous energies would unavoidably collide with photons of microwave radiation left as an echo of the Big Bang. These collisions, it was believed, would annihilate all cosmic-ray particles with energies above about 10^{19} electron volts.

But experiments have proved this wrong.

"So the mystery of these cosmic rays deepens," Dr. Cronin said. "But what better science can you have than when the mystery thickens?"

—M.W.B.

Jimmie Holland

Listening to the Emotional Needs of Cancer Patients

Dith Pran/The New York Times

For a time in the early 1970's, conversations between Dr. Jimmie Holland, a leading psychiatrist, and her husband, Dr. James Holland, a leading cancer specialist, often took a familiar course:

Dr. Holland the oncologist would relay exciting data coming out of the country's first cooperative cancer trial, statistics about white cells and average survival and tumor size.

Dr. Holland the psychiatrist would listen and then ask a question that stumped her husband: "Yes, but how do the patients feel?"

"He had to answer 'we don't know,' since there had been virtually no work in the area," she recently recalled, speaking with a trace of a drawl from her native rural Texas where, she said, "little girls are named Bobbie, Billie and Jimmie."

Since then Dr. Jimmie Holland has worked with cancer patients, parlaying that quintessential psychiatrist's question—"Yes, but how do they feel?"—into the first psychiatry service at Memorial Sloan-Kettering Cancer Center, the country's largest training program in psychiatric oncology and, ultimately, into the emergence of psycho-oncology as a new field.

For the past 17 years she has conducted tireless research about how battles with cancer affect the mind: how does body image change after the removal of a breast or testicle? What is the best way to treat depression in patients undergoing cancer treatment and to treat anxiety in those who have survived? How do you measure the quality of life after different cancer cures, so patients treated with aggressive therapy will emerge glad that they're alive?

Along the way she has helped thousands cope: patients, their families, their doctors, even their pets. On a recent afternoon of rounds with her staff, she planned treatment for a young woman with leukemia who was suffering from panic attacks, a man with lung cancer who had decided to be taken off a ventilator so he might die, a middle-aged woman being treated for ovarian cancer who had returned to work after a bout of depression.

"We still get a lot of people saying: 'Of course she's depressed. She's got cancer. Who wouldn't be?' " she lamented. "But we have learned that most people with cancer are sad but don't have true depression. And when they do, it can be treated."

Twenty years ago, the prognosis for cancer patients was so grim that the few therapists who worked with them focused solely on death and dying. Patients who survived a few years were regarded as physiologic miracles, bereft of psychological needs.

"The patients would say, 'Doctor, I've lost all my ability to have sex,' and the doctor would say, 'You're lucky to be alive,' " Dr. Holland recalled.

But advances in the treatment of cancer and a growing consumer awareness forced cancer specialists to pay attention to their patients' emotional lives.

"Suddenly there were all these survivors and no one knew much about their emotional needs," she said. "The 70's were clearly the right time to study the psychological impact of cancer and we just rode the crest of that wave."

It is a modest self-evaluation from a woman whose 65 years have included some unusual twists and turns. An only child, born to parents who never finished high school, Jimmie Coker grew up on a farm in Nevada, Texas, where she loved taking care of animals. When she graduated from high school—in a class of eight—she planned to become a nurse. "But then I figured out I could be a doctor," she recalled.

After graduating from Baylor University in Waco, she went to Baylor's medical school in Houston, where her class of 80 had "three women, three Jews—the usual quota in Texas" at that time, she said. Although she initially planned to take care of patients' physical problems, she quickly shifted course. "I decided the most interesting thing was not the biochemistry of congestive heart failure, but the psychological and social issues surrounding it," she said. "How do people cope with adversities in life? The best common denominator to study this is illness." Although she initially worked with patients suffering from diverse diseases, after her marriage, in 1956, her interest gradually shifted to the psychology of cancer.

Dr. Holland listened to her oncologist husband talk about the emotional problems of cancer patients for years, as she worked part time while raising their five children. When she returned to full-time work in the mid-1970's, she knew where to focus. In 1977 she and two of her students started the division of psychiatry at Memorial Sloan-Kettering Cancer Center, one of the nation's leading cancer hospitals, which had previously had only a half-time psychologist on staff.

"When I started, I was in a strange foreign environment," she recalled. "Few of the medical staff came around. First I had to prove to the oncologists that we could help their patients—before they were ready to go to a psychiatric hospital or jump off the roof. But once we did, our work just began to grow."

Dr. Holland and her team had to build the field from the ground up, developing methods for diagnosing and treating psychiatric problems in people with cancer as well as for measuring the effects of various treatments on their subsequent quality of life. They had to determine, for instance, how common depression was in patients with different types of cancers. And they had to find out which psychiatric drugs would not interfere with chemotherapy.

They have since determined that about a quarter of the people with cancer have significant psychiatric problems, although many patients do not get help. Dr. Holland estimates that her staff now sees about 10 percent of all inpatients at Memorial Hospital, some referred by their medical doctors or families and some who initiate the contact themselves.

And the frequency of psychiatric problems is much higher than 25 percent in some groups.

"Adolescents are already going through struggles with their identity, and getting cancer has a damaging effect," she said. "They are almost always worried about how will they be accepted back in school, and we are trying to improve their re-entry." Dr. Holland and her colleagues are completing a study of adults who survived cancer as adolescents 25 years ago to see how it has affected their lives.

They have also devoted special attention to patients undergoing particularly uncomfortable forms of treatment, like bone-marrow transplants, because that therapy can involve prolonged pain and suffering and patients must remain in isolation rooms for weeks or months. Before treatment, the therapists frequently evaluate patients to make sure they are emotionally strong enough to weather the course, and they monitor all patients during treatment.

Studies have shown that the most common problems during cancer treatment are anxiety, fear and depression. More surprisingly, they have also found that subtle depression and anxiety are common after therapy is finished, even when the patients appear to be cured.

"Most people think of themselves as invulnerable, but people who have had cancer cannot think like that," Dr. Holland said. "Many feel vulnerable, like a sword of Damocles is hanging over their heads. They have the sense that they are damaged goods. They worry, 'If I want to get married, am I whole?' They have trouble with intimacy.

"You'd expect people to feel good when they finish treatment. But from their perspective, the medicine that has kept their disease at bay has stopped and they are not going to see their doctor, who has been their ally in fighting the disease every week." At Dr. Holland's urging, Memorial has set up a "post-treatment" center to meet the emotional needs of cancer survivors.

Although Dr. Holland and her staff use many of the same weapons as other therapists—drugs, counseling and sometimes electroshock therapy—they have had to adapt their tactics for their unusual patient population.

Some may be too sick or weakened by ongoing treatment to sit through a 45-minute therapy session, so short, frequent appointments are substituted. Others may be intermittently homebound, so telephone sessions—or conversations via videophone—may be interspersed with face-to-face consultations. Also, patients frequently broach topics beyond the scope of usual therapy.

"You are treating someone who may be facing death, so you tend toward more existential questions: what is life and what is death," said Dr. Holland, noting that such questions make even many psychiatrists uncomfortable.

Dr. Holland has a history of tackling subjects that others are reluctant to touch.

In 1972, at the height of the Brezhnev regime, she and her husband moved to Moscow for a year so he could study chemotherapy and she could study how Soviet doctors defined "schizophrenia," a diagnosis then sometimes used by the Soviets to imprison dissidents. She found that while doctors in both countries used broad definitions of the disease and treated it with similar kinds of drugs, those in the Soviet Union were far more likely to use the "big lock-and-key approach."

"They didn't tolerate much aberration on the streets of Moscow in those days," she said. "They had a diagnosis called 'sluggish schizophrenia,' which involved delusions that you could change the world, even change the system—that was where the dissidents fit in." Still, she regrets that the image of Soviet psychiatry has been tarnished by being used for political ends, noting that only a very few psychiatrists—and none of the ones she worked with—were responsible for the abuse.

Although Dr. Holland is gratified to see that more attention is being paid to the emotional lives of cancer patients, she said there is a need for more good science in the field because studies often contradict the popular wisdom.

For example, although many doctors have assumed that a woman with breast cancer would be happiest if she received treatment that would preserve the breast, rather than undergoing a mastectomy, Dr. Holland said that is not always the case.

"The studies that are most useful have shown that women who do best emotionally are the people who get what they want," she said. "Some people say, 'Offending organ, get thee gone; I don't want an organ with cancer.' Others say, 'My breast is very important to me.' "

And Dr. Holland absolutely bridles at the popular notion that patients must think positively and fight the disease to be cured.

"Patients get bombarded by family members who say, 'You aren't trying hard enough,' " she said. "I think we're overdoing it, trying to make everyone into a warrior. Some people are not good at it. Each of us has a different style of coping, and many work equally well."

After two decades of studying how cancer treatment affects the mind, Dr. Holland and her staff are now beginning to study the effect of the mind on cancer treatment.

She said there is evidence that the body's physiology may be altered by the intense anxiety experienced by patients at high risk for cancer—such as women with close relatives who have died of breast cancer—and by cancer patients facing treatment. For example, patients on weekly chemotherapy frequently develop nausea and mild immune-system dysfunction even before the drugs are administered. Dr. Holland wonders if these side effects could be reduced by treating the anxiety.

In addition, by dint of place and profession, Dr. Holland and her staff have been predictably drawn into the debate over physician-assisted suicide, a practice she opposes.

"We get requests for help with committing suicide, but it's really a cry for help. If you treat people, set them up with home care, give them the sense that someone cares, their position usually changes."

—ELISABETH ROSENTHAL
July 1993

A HAPPIER MIND FOR A HEALTHIER BODY

Nearly three decades after Dr. Jimmie Holland began her research, few cancer doctors would dispute the importance of paying attention to the minds of their patients as well as their bodies.

But in recent years, researchers have begun to discover that treating cancer patients' depression and anxiety may do more than help them cope with illness: in some cases, it may help them live longer, too.

A growing number of studies, said Dr. David Spiegel, professor and associate chairman of psychiatry at the Stanford University School of Medicine, suggest that "depression and hopelessness are associated with a poorer outcome."

Of 24 published studies examining the link between depression and the progression of disease in cancer patients, 15 have found that patients who were severely depressed suffered relapses earlier than nondepressed patients, or survived for a shorter period of time.

Depression may be a risk factor for developing cancer as well. Though the evidence is mixed, five studies found that people who suffered from chronic and severe depression were more likely to fall ill with the disease.

Conversely, offering patients tools to help them deal with the emotional impact of their illness may prolong survival, some studies indicate.

In a 1989 study by Dr. Spiegel and his colleagues of 86 women with metastatic breast cancer, the patients who participated in weekly sessions of group therapy lived an average of 1.5 years longer than those who did not participate. Forty-eight months after the study began, Dr. Spiegel said, all the women in the control group had died, while one-third of the women who received the therapy were still alive.

Two other studies, one with melanoma patients and the other with patients suffering from lymphoma or leukemia, found similar differences in survival rates.

The results of such studies are not conclusive: some other researchers have failed to reproduce the findings, and at least three other studies on the issue are not yet completed.

Still, Dr. Spiegel said, the consensus among researchers is that asking whether psychological state can affect how long a patient lives "is a fair question."

He and other researchers, however, cautioned against interpreting such findings as support for the notion that having a positive attitude can make a difference in cancer survival.

"There's a New Age counterculture idea that if you have the right attitude it will go away, and if you have the wrong attitude, it won't," Dr. Spiegel said. "But all that does is burden people with guilt."

In fact, a 1999 study by Dr. Margaret Watson and her colleagues at the Royal Marsden Hospital in Sutton, England, found that breast cancer patients who demonstrated optimism and a "fighting attitude" toward their illness survived no longer than women with a less positive attitude. The study, which followed 578 women with early stage breast cancer over five years, also showed that subjects who found it hard to express negative emotions were no more likely to relapse or to die.

Optimism is important, said Dr. Watson in an interview after the study appeared in the medical journal *The Lancet,* because it helps women feel in control of their illness. "But many women who have fought will very likely have times when they don't feel like it, and it's helpful to relieve them of the burden of guilt," she said.

Dr. Spiegel agreed, saying that the goal of studies like his "is to make our treatment of patients more comprehensive, not to blame them for being sick."

Depression, for example, is still often overlooked by doctors. Yet even if hopelessness and a depressed outlook do not turn out to hamper survival, they take their toll. Depressed patients have more trouble adhering to treatment, use medical care more, are more difficult to treat because of their anxieties and do not do as well medically.

Given this, spotting the symptoms of depression early and providing treatment quickly can only be beneficial, Dr. Spiegel said.

"I think Jimmie would say that, too," he added.

—Erica Goode

Joe Z. Tsien

Of Smart Mice
and the Man Who
Made Them That Way

Laura Pedrick for The New York Times

A certain amount of disorder has broken out around Dr. Joe Z. Tsien, a neurophysiologist who just announced that he had created a smarter strain of mouse by genetically altering a gene for memory. Patients call seeking help. Individuals of enhanced imaginations warn that the mice may escape and take over the planet. Television crews patrol the halls. His voice-mail box has overflowed.

But Dr. Tsien, seemingly the only scientist on the Princeton campus who, on a warm summer day, is wearing a tie, ignores the chaos and a phone that rings every couple of minutes. In soft tones he describes the remarkable journey that has led him from Wuxi, a small town near Shanghai, to the position of having made a significant, maybe decisive, contribution to understanding the nature of memory and intelligence.

Dr. Tsien (pronounced chee-YEN) says he did not begin to consider the wider implications of his work until just before his article was published (*Science,* September 2, 1999). He engineered his smarter mice for purely academic reasons, to address and perhaps solve the question of how memories are laid down in the brain. But the mice turned out to be smarter as well as having better memories, lending an unexpected new dimension to the experiment.

Although many arguments with psychologists doubtless lie ahead, Dr. Tsien believes that learning, memory and intelligence are all intimately related because, as his smarter mice demonstrate, "a common unifying mechanism underlies them all."

And because mice and people use the same basic mechanism of memory, the smarter mice could well shed much light on the nature of human intelligence.

Dr. Tsien's result, as he is the first to note, rests on knowledge and techniques developed by other scientists. He describes his experiment as "obvious"—at least in retrospect. His achievement lies in the fact that, in a highly competitive field of biology, he was the first to conceive of the experiment and to see that it could be decisive. He also carried it out in a particularly convincing way. "Extremely nicely done," was the verdict of Dr. Eric R. Kandel, a leading biologist at Columbia University and the former laboratory chief of Dr. Tsien.

The idea that led to the smarter mice was no lucky break. Rather, it was a feat for which Dr. Tsien had been preparing intensively for many years, including seven years of postdoctoral education. In Wuxi, where his father was a clerk and his mother an accountant, he was the only person to enter college from his high school, one attached to a fabric plant. But the college was a good one, the East China Normal University in Shanghai, and he decided to do doctoral studies in the United States.

"In 1986, China was still very closed, so we really had no idea about the United States," Dr. Tsien says in describing how he picked a college. He chose the University of Minnesota because it offered to waive the application fee, which he could not afford, and because the Chinese characters for Minnesota translated invitingly to "clean air blue sky."

Once he recovered from the surprise of finding the clean-air-blue-sky state so cold, he developed an interest in neurophysiology and the instruments then available for monitoring the electrical signals transmitted by brain cells. "I got fascinated by seeing a nerve cell fire. They are talking—what does that mean?" he says.

A long apprenticeship was necessary before he could begin to parse that language. He did his Ph.D. thesis with Dr. Lester R. Drewes of the University of Minnesota, helping him conduct studies under a Defense Department grant on how the warfare agent sarin blocks the transmission of nervous signals.

Receiving his Ph.D. in 1990, Dr. Tsien was accepted as a postdoctoral student by Dr. Kandel's laboratory. There he worked on identifying genes that are active in rats' brains during memory formation. "I got a more systematic education in neuroscience," Dr. Tsien said. "I got to see how a big lab operates."

He then moved to another leading neuroscience laboratory, that of Dr. Susumu Tonegawa at the Massachusetts Institute of Technology. Dr. Tonegawa won the Nobel Prize in physiology or medicine in 1987 for research on the genetic control of the immune system, and later switched to the study of learning.

In Dr. Tonegawa's lab, Dr. Tsien worked with so-called knock-out mice, animals from which a gene has been deleted. The idea is to learn what a gene does by excising it and seeing what defects the mouse develops. He became interested in the brain cell component, known as the NMDA receptor, suspected of being central to the memory mechanism. The receptor consists of parts made by several genes, the chief part being specified by a gene called NR1.

Using advanced genetic techniques, he decided to create a mouse lacking the NR1 gene in the cells of its forebrain. Creating the mouse took two and a half years and, for a postdoctoral student, was a substantial risk. If the experiment failed, there would be no result worth publishing.

In the end, he was able to knock out the gene in just the cells of the hippocampus, a brain module dedicated to learning and much studied by neuroscientists. "I think a god looked on me very kindly," he said, referring to the element of luck in creating such a valuable research tool.

The mice lacking the NR1 gene in the hippocampus indeed did not remember as well, suggesting the NMDA receptor is important in laying down memories. But the experiment, published in December 1996, was regarded by other experts as less than fully conclusive, because the absence of the NR1 gene could have caused general brain damage not specific to memory.

Dr. Tonegawa became very interested in the mouse, as did Dr. Kandel, because Dr. Tsien had made it with a technique developed in Dr. Kandel's laboratory. The two lab chiefs were also interested in receiving due credit, and discussions ensued between them that were stressful for him, Dr. Tsien recalls.

However, he now had sufficient credentials to set up his own laboratory. "After working with these two powerful people, I wanted to be free," Dr. Tsien says. Two years ago he was appointed an assistant professor at Princeton and was able to set up his own lab. He began to think about how he might try to improve a mouse's memory, rather than sabotage it, because such an experiment would run far less risk of being criticized as nonspecific.

The focus of his thinking was the anatomy of the NMDA receptor. The intricate biological device is shaped like a cylinder or doughnut embedded in the outer wall of certain brain cells. Usually its central channel is firmly closed. But when the nerve cell receives signals from two other nerve cells at the same time, the NMDA channel springs open, allowing a current to flow into the cell. This current generates a long-lasting change within the cell, making it much more responsive the next time that either of the two other nerve cells is active alone.

This property of the NMDA receptor—opening when two signals arrive simultaneously—has long been suspected to be the basic mechanism of memory, because it is a way for the brain to make an association between two events. The exact degree of simultaneity turns out to be very important. In young mice, two signals can arrive as much as one-tenth of a second apart for their coincidence to change the nerve cell. In older animals, the NMDA receptor allows a much narrower window of time for an association to register.

Another known fact about the receptor was that its composition changed with the age of the animal. Its main component is the gene product of NR1, which Dr. Tsien had knocked out in his memory-deficient mice. But the NR1 component works with any of four different partners, which modulate its activity in different ways. Two of the partners, known as NR2A and NR2B, are particularly important in cells of the forebrain. As animals age, there is a switch from NR2B to NR2A as the preferred partner for NR1.

Abilities in many animals decline after sexual maturity. Songbirds cannot learn new songs. The human mind becomes less flexible at learning new languages. "I am always stuck with my Chinese accent, but if I had come to the United States 12 years earlier I would have learned perfect American," Dr. Tsien says by way of personal example.

As he contemplated these various pieces of information, Dr. Tsien said, it seemed clear that they were related. The natural switch with age of NR1's partner must underlie the increasingly stringent requirement for two signals to arrive simultaneously, and the narrowed window of time must be the reason why older people find it harder to make associations.

But no one had specifically stated the idea in those terms, as far as Dr. Tsien knew. And certainly no one had done the obvious experiment, which was to engineer mice in which the NR2B gene was artificially put into hyperdrive to see if their memories improved.

So Dr. Tsien took a copy of the mouse NR2B gene and linked it to a special piece of DNA, called a promoter, that is active only in cells of the mouse forebrain. He injected this genetic fragment into fertilized mouse eggs, where it added itself to the mouse's normal complement of genes. Because of the promoter, the NR2B gene was active in cells of the forebrain, adding its product to that produced by the mouse's own NR2B gene.

With all the extra NR2B being produced in the mice's brain cells, the NMDA receptors underwent a subtle but significant change. Instead of staying open for 100 thousandths of a second, as they do in normal mice, the receptors' interval increased to 250 thousandths of a second.

That minute biophysical change, Dr. Tsien says, is what underlies the superior learning skills of the mice. The essence of smartness is an extra 150 thousandths of a second.

Dr. Tsien is waiting to see what the public and his peers in the neuroscience community make of the implications of creating smarter mice. "To the scientific community this is a small step for a man," he says. "The fundamental question is, 'Is this a giant step for mankind?' "

What does he think? "I don't know," Dr. Tsien replies.

—NICHOLAS WADE
September 1999

OUTSMARTING NATURE

Has nature designed the human body to be as good as possible? Evolution's driving motif—the survival of only the fittest—suggests the answer is yes: perfection of some sort has been achieved. But whatever we may be optimized for is some goal of nature's, and not necessarily one that we would choose our-selves. Otherwise we would all be supremely beautiful, as smart as heck, and live for 1,000 years.

Dr. Joe Tsien put his finger on an interesting piece of this puzzle by geneti-cally engineering a strain of smarter mice. So now we know: mice are not as smart as they could be, and the same almost certainly goes for people too. Whether there are hidden trade-offs in greater smartness remains to be seen.

Dr. Tsien did not set out to demonstrate that people could be smarter than they are, though that was a principal reason for the considerable interest his jour-nal article aroused. His experiment was designed to test a fundamental proposi-tion about how memory works at the level of the individual nerve cell, or neuron. His discovery has wide implications for medicine, particularly diseases of memory.

For many years the principal way of studying the brain has been by sticking sensitive needles into neurons and recording the pattern of electrical activity. New methods of scanning the living brain have helped visualize the behavior of large groups of neurons as the brain performs various tasks. But the most promising source of new progress may lie with the techniques that Dr. Tsien has mastered, those of manipulating the various genes thought to be involved in the neuron's operations.

The genes, after all, translate the blueprints that contain the specifications for the brain's architecture and operation. Figuring out how these genes work offers a bottom-up approach to exploring the mechanics of this extraordinary machine.

Dr. Tsien works by adding or deleting genes from mice. This is a powerful experimental approach because although the mouse brain is far simpler than the human brain, its neurons and much of the basic architecture are probably quite similar. The comparative simplicity of the mouse brain is of course an advantage, because it will presumably be easier to understand. Mice can also be genetically manipulated in ways that would be unethical in people.

Although Dr. Tsien is interested in the academic goal of understanding how the brain works, his line of inquiry is likely to spin off some interesting practical possibilities. If a pharmaceutical company were to develop a drug with the same effect as the gene that makes mice smarter, maybe people could take it during job interviews, when taking exams or when trying to fill out IRS forms. Maybe the GNP could be increased if everyone got smarter. Such pleasing speculations, whether well founded or not, give a glimpse of how important the work of Dr. Tsien and others could turn out to be.

—N.W.

High Seas Hunter Pleads for Preservation of Fish

Courtesy of Dr. Carl Safina

Dr. Carl Safina was edgy, drumming his fingers, glaring at the sea through dark glasses. His boat, the *First Light,* rolled in the low waves. Seawater sloshed and gurgled through a chum bucket, forming a trail of chopped-up fish and an oily slick behind the boat.

Dr. Safina scanned the horizon and glanced at his fishing rods. Their lines trailed off into the slick. Somewhere below were three-inch steel hooks, which he had set with chunks of bloody fish many hours earlier.

The sea off the eastern tip of Long Island once boiled with sharks. The fury provided a semifactual basis for "Jaws" and a mythology about sailors and swimmers threatened by fierce predators with razor-sharp teeth.

But today, the sharks are fewer, and catching one of the killers can frustrate even professionals like Dr. Safina, a marine conservationist who clearly loves the hunt but is haunted by a vision of fishery collapse. The danger was driven home to him long ago when he overheard a conversation in which a fisherman vowed to catch every last tuna because no one had left him any buffalo. Alarmed, Dr. Safina became determined to do something about it and in the process became one of the world's leading exponents of fishery restraint, a potentially lofty stance tempered by his deep joy of the catch.

"Some days it's unbelievable," he said wistfully. "You run into fish wherever you go. Other days, it's like this, dead."

A dorsal fin appeared and moved back and forth slowly. Another came closer and disappeared.

Dr. Safina, moving quickly and silently in old sneakers, picked up a rod and began to reel it in.

Suddenly the the rod bent down and the reel whined.

"I've got one!" he cried.

Eleven minutes later, after a frenzy of tugging and splashing, a blue shark more than seven feet long darted away from the boat, freed purposefully. The newly liberated fish bore a small metal tag stamped with an identifying number: 234474.

Dr. Safina is a seafood lover. His widely praised book, *Song for the Blue Ocean* (Henry Holt, 1997), a survey of global fishery woes based on his travels, is dotted with depictions of restaurants, of sushi bars, of business done over plates of grouper and giant prawns.

He also admits to liking the taste of shark, including a big mako he once fought for more than two hours before landing the exhausted, 232-pound creature. ("It was incredibly good," he recalled.)

But times are changing, he said. Many varieties of sea life have been dangerously thinned by overfishing, the monumental bad habit of a global, multibillion-dollar industry ranging from mom-and-pop boats to factory trawlers. Sharks are caught in the upheaval, he said. They mature late and bear relatively few young, and scientists fear that whole populations are under fire and might not recover for decades, if ever.

"I love the hunt and know the thrill of the kill," he said, still elated from his catch. "But I'm not sure we should be doing it. They need a break."

Tagging is another matter. Dr. Safina said the process, fun for sportsmen and easy on the sharks, aids a Federal program that tracks the big predators, which turn out to migrate surprisingly far, sometimes thousands of miles.

To a striking degree, Dr. Safina, 43, has succeeded at turning his ideas into action—not only in tagging sharks off Montauk on a summer day, but in pioneering global fishery conservation measures. Articulate and energetic, knowledgeable and willing to take a stand, he is something of an eco-warrior ready to fight public indifference and lax regulators. His role has already won him many friends and foes.

Enemies include commercial fishermen who set long lines that can stretch for as far as 60 miles, as well as those who hunt bluefin tuna, a giant so coveted as seafood that its breeding stocks in some areas are seriously depleted. A single bluefin once sold for $83,500.

An ecologist, Dr. Safina has worked since 1979 for the National Audubon Society. In 1990, he founded its Living Oceans marine conservation program, which he runs. From that post he has agitated widely, teaching at Yale, writing for *Scientific American,* testifying before Congress, serving on Government fishery panels as well as international bodies. From 1991 to 1994, he served on the Mid-Atlantic Fisheries Management Council of the Department of Commerce, appointed by the Secretary.

His wins are impressive. After he formally raised the issue of declining shark populations with the world group that monitors trade in endangered species, it agreed in 1994 to collect data on sharks—the first time it did so for a big-money fishery. In 1995 he was a force behind the United Nations agreement to adopt a new global standard by which fishermen on the high seas must err on the side of caution when it comes to questions about how far fish populations have fallen. In 1996, Congress adopted some of his ideas in the sustainable fisheries act, which bans practices in United States waters that quickly deplete stocks.

Early in 1998, he worked with several environmental groups to get swordfish off the menus at top restaurants around the country. Commercial fishermen call such acts extreme, saying catches have already been reduced enough to aid the fish's recovery.

Of late, Dr. Safina has focused on sharks. A recent report by his Audubon team said that in the past two decades populations of large coastal sharks in the Atlantic and Gulf of Mexico had dropped as much as 85 percent and were in danger of declining further unless states in these regions began to enact and enforce protective regulations. In its wake, a number of states said they would review and possibly tighten rules.

A hallmark of Dr. Safina's activism is depth and breadth. In researching his book, he traveled to 10 countries on four continents. In Hong Kong, he found streets lined with shops selling shark fins, prized for promoting health and vigor.

"The fractional remains of hundreds of thousands of sharks must be moving through these markets," he mused in his book. Conversations with shopkeepers suggested that many of the sharks came from the Montauk area.

Now, in late July, he had returned to these waters, to an area he had known intimately over the years and had worried about as catches of all kinds of fish had declined. His dark hair, longish and streaked with gray, curled out from beneath a cap. He was deeply tanned and wore no watch. Skewering a squid through a big hook, he cast the bait into the gently rolling water.

"I was seeing things in a way I wanted to communicate," he said of his reasons for writing a book. "I wanted lots of people to care, so that things had a chance to change."

He paused. The engine and radio were off. The only sounds were the gurgle of the chum bucket and the lapping of low waves.

"Partly," he added about his book, "I wanted people to feel a sense of outrage, to know that the people being paid to guard the hen house are foxes."

Referring to fishery regulators, who set quotas and are often seen as cozy with big business, he said: "They mostly got better than they deserved. There was much worse that could have been said."

Carl Safina was born on May 23, 1955, to a middle-class family in Brooklyn. When he was 10, his family moved to suburban Long Island, and his father took him repeatedly to its pebbly north shore to hunt striped bass.

"I thought that my father and I, with our secret fishing spot, were very, very lucky," he recalled in his book. "The arrival of the bulldozers on our shoreline was an unfathomable catastrophe. In a youngster's worldview, the country map is small. Loss of my beach amounted to expatriation. And no one is so long-remembering as a refugee."

As an adult, he watched as ripples of development spread. Eventually, they moved through the distant fishing grounds he had come to love off Long Island. In growing alarm, he studied ecology throughout the 1980's, earning a doctorate from Rutgers in 1987.

A turning point came in 1989 while he was fishing about 50 miles south of Fire Island. The bluefin tuna were thick, and armadas of fishermen were doing everything in their power to catch the shimmering gold.

"People were getting ridiculous amounts," he recalled. "Somebody got on the radio and said, 'Guys, maybe we should leave some for tomorrow.' Another guy came on and said, 'Hey, they didn't leave any buffalo for me.'"

That image came to haunt Dr. Safina. In his work and writings, he began to refer to global overfishing as "the last buffalo hunt," a phrase that eventually joined the lexicon of marine conservation.

In the 1990's, he dedicated himself to reversing the trend, mainly by using his scientific and persuasive skills in the field of fishery regulation and management. While often succeeding, Dr. Safina had plenty of frustrations and setbacks.

In summer 1993, he set aside his routine by leaving Islip, where he lived and worked, and renting a room at the Montauk marina where he kept his boat. There he meditated. Day and night he pored over notes and ideas, occasionally taking a break out at sea. Eventually he hit on a book plan that excited him.

"I wanted to preach outside the choir," he recalled.

A labor of love, the book recounts his travels to 18 shorelines around the globe and profiles ongoing battles between fish and people. The sites, which double as chapter titles, include Ogunquit in Maine, Yachats in Oregon, Malakal in Palau and Sulu in the Philippines.

In Sulu, he wrote, he stumbled into armed battles between local warlords fighting for political dominance, and at times feared for his life as he investigated how communities of poor fisherman are trying to kick the deadly habit of poisoning coral reefs. When squirted into reefs, sodium cyanide—the same stuff used to kill prisoners—stuns fish and allows easy catches, mainly for the aquarium trade. But corals subsequently die.

The Pacific is widely scarred by such poisoning. But in Sulu, a tropic archipelago that forms the southern flank of the Philippines, some local fishermen are trying to unlearn self-defeating chemical tricks and instead make a living with nets and traps. Dr. Safina dived with the men on a reef scarred by cyanide poisoning and dynamite blasts.

"One trainee works for 10 minutes to entrap a single emperor angelfish," he wrote. "He finally succeeds, carefully removing the struggling blue booty from the net and slipping it into his mesh-topped bucket. This little emperor is worth about $12 to him. Someone in the U.S. will pay about $150 to acquire it."

The book is often melancholy. Dr. Safina said its title, *Song for the Blue Ocean*, reflects not only the sea's obvious color but its not-so-obvious state of decline. Despite the blues, he added, his own song for the ocean contains hints of optimism. One refrain is that the sea has huge regenerative powers and responds generously whenever people curb their appetites. He ticked off recent successes in fisheries restraint, crediting the increasingly large and feisty alliance of environmental groups and sports fishermen. "We've accomplished some real long shots that we had every reason to be pessimistic about," he said.

Suddenly, after an extended lull, a large school of skipjacks broke the surface, offering a tantalizing opportunity for Dr. Safina, the sport fisherman, raw and undisguised. He had no tags for small game. He grabbed his tuna rod and cast hard, the bright lure falling into the agitated water that marked the school's leading edge.

"Fish on," he said with a grin as the rod bent over.

He reeled in a glistening, 20-inch skipjack, its sides a rainbow of colors, its back electric blue.

"Very pretty," he said.

While still in the water beside the boat, the handsome fish slipped off the hook and sped away.

Dr. Safina looked on contentedly. Earlier, he had used a file to remove the hook's barb as part of his inconspicuous war on the mentality of excess.

—WILLIAM J. BROAD
September 1998

TAKING IT TO THE CONSUMER

In his newest effort to save sea life, Carl Safina is aiming at the most ferocious predator at work in the world's oceans—people. As overfishing thins the ranks of prized species like cod, tuna and swordfish, Dr. Safina and his group, the Living Oceans marine conservation program of the National Audubon Society, are urging consumers to choose less popular species in an attempt to help repair damaged ecosystems and to support sustainable fisheries. Their lists recommend seafoods to eat and avoid. Making wise choices in buying seafood, Dr. Safina said, is a new way to help assure healthy oceans for the future.

"Fisheries management is in many cases deadlocked by politics and industry lobbying, resulting in depleted fisheries," he said. "Consumer power is potentially enormous in moving fisheries toward better health."

Based in Islip, New York, the Living Oceans Program first published a guide to seafood in 1998. It made recommendations among 21 species, and later expanded the list to include 34 species. The ratings ranged from green to red—

from fine for dining (tilapia, striped bass, mahi-mahi) to better left uneaten (snapper, grouper, Atlantic halibut).

"As you can imagine, not everybody in the red category was pleased with us," Dr. Safina said of fishermen who saw their seafood products recommended for a consumer thumbs down.

Criteria for rejection included evidence of population declines, habitat destruction, pollution risk and large bycatches—the unwanted creatures that fishermen often catch inadvertently by the ton and then throw back into the water, usually dead.

More recently, Dr. Safina and the Living Oceans Program have published the Audubon Seafood Lover's Almanac, a 96-page guide that expands on the advice and tries to emphasize the positive. "Bluefish," the almanac says, "is one of the very few fish for which fishing controls were put in place before problems began." On the same page, it gives a bluefish recipe, Lady M's Marinade, which is laced with ginger, parsley, oregano and basil.

Taking his efforts at consumer education even further afield, Dr. Safina has joined the advisory board of a new company in New Hampshire that sells what it calls sustainable seafood from healthy fisheries. EcoFish, which does its business mainly on the web at www.ecofish.com, says it donates a quarter of its pre-tax profits to marine conservation.

Dr. Safina believes consumers are willing to ease up their demand for depleted and threatened fish stocks—particularly if they can get seafood from well-managed, conservation-minded fisheries. As he says on the EcoFish web site, "Before consumers can vote with their wallets they need to know how to find fish caught in more sustainable ways."

—W.J.B.

Mary-Claire King

Quest for Genes and
Lost Children

Marty Katz for The New York Times

Mary-Claire King, a geneticist of international renown and neutronic stores of energy, is sitting in her sunny, unprepossessing office at the University of California in Berkeley, talking about the current monarch of her many passions. She is trying to find the gene for hereditary breast cancer, a gene that could be of great significance to hundreds of thousands of women in the United States who are at great risk for early onset of the disease. She has been seeking the gene for 17 years, weathering the skepticism of her colleagues, and even sometimes her own doubts.

Not long ago she found the approximate location of the gene, and now she and her students are homing in on the trophy proper. She wants it very, very badly, and she believes her laboratory is very, very close.

She also knows that other labs have since joined the race, and she would hate to see some newcomer step in at the final hour and seize victory.

"It could be in there right now, sitting on one of our plates," she says, referring to the petri dishes where segments of isolated genetic material await analysis. Her voice intensifies, and her deep dimples disappear along with her smile. "We're obsessed with finding the gene," she says. "I want it to happen in our lab."

One of her students pokes his head in the door, grinning broadly, and says he has something to tell her. She excuses herself and joins him in the room next door. Suddenly, a loud crow of delight fills the halls: "Yes! Oh, yes! That's WONDERFUL!" She returns to the office, her face glowing. Has the gene been found? Are the scientists even nearer their goal than she suspected? "He just told me he's getting married," she says. "I am so, so happy for him."

That Dr. King should react with untethered joy to her student's joy is hardly surprising. Though she was trained as a mathematician and is now a molecular geneticist as committed as any basic researcher to rigor and abstraction, nearly everything she has ever chosen to work on has had, at its core, a deep sense of humanity. She won her greatest fame by working in Argentina with a human rights group, the Grandmothers of Plaza de Mayo, attempting to reunite with their families children who were kidnapped in the 1970's and early 1980's by the Argentinian military junta.

By analyzing genetic material from the children and comparing it to the genes of grandmothers and other relatives who survived Argentina's eight-year "dirty war," Dr. King and her co-workers were able to prove that many children had been snatched away as infants and given to other families, while their real parents were either shot outright or mysteriously disappeared.

Dr. King also has immersed herself in the case of El Mozote, a village in El Salvador where, in 1981, at least 794 peasants, many of them children, were massacred by American-trained soldiers of the Salvadoran military. The first skeletons of the victims were dug up last October, and the Government of El Salvador has agreed to permit a thorough forensic analysis of the remains once the exhumations are complete. Dr. King and other researchers have already begun trying to identify the skeletons by comparing DNA extracted from bone and teeth with that of living relatives, for possible use in criminal proceedings.

"We'll be working with the UN, with the same forensics team that we did in Argentina," said Dr. King. "But this will be much, much more difficult, because there are very few survivors," and thus nothing to compare the DNA to.

Dr. King, who is 47, is an unreclaimed liberal, and she is delighted that her headquarters, wedged in the middle of a building devoted to forestry science, has a bit of history to it. In 1970 in that very room, as a graduate student at Berkeley, she and other students organized a letter-writing campaign to protest the American invasion of Cambodia. They gathered 30,000 signatures from voters in northern California.

"We started to rally together the same forces when the Persian Gulf crisis hit," she said. "But it didn't turn out to be necessary to do anything."

Yet Dr. King is not given to political posturing, and she is amused to say that she is now collaborating with the United States Army. "We're working with our own Government, I pale to say it, on MIA cases," she said, including an attempt to identify a man shot down in a fighter plane during World War II whose body was preserved in the bog where it landed. The King lab does not consider itself a forensics lab per se, but its researchers have perfected a means for extracting DNA from teeth, taking it from the nerve pulp that remains. And teeth, it turns out, are better preservers of genetic material than are bones.

Dr. King is also enough of a pragmatist to have hoisted herself up to the summit of mainstream science. She was a strong candidate to replace Dr. James Watson as the director of the Human Genome Project, the celebrated Federal enterprise to map and analyze all 100,000 human genes; the job went to Dr. Francis Collins, a leading geneticist at the University of Michigan, who is collaborating with her on the quest for the breast cancer gene. Dr. King was asked to apply for the job as head of the National Institutes of Health, to replace the departing Bernadine Healy, but declined to be considered.

"I'm not interested in a job with that level of administrative responsibility," she said. "It would be too far removed from what I love to do, which is science."

Yet as pure scientists go, Dr. King has a pronounced bedside bent. She and two other researchers just published a report in *The Journal of the*

American Medical Association, anticipating the upcoming isolation of the gene behind early-onset breast cancer, and discussing the possible options for those who carry the mutant gene, an estimated 600,000 women in the United States. Such women are at extremely high risk of contracting breast cancer before the age of 50, the researchers said, and they must think carefully about whether to take such drastic measures as having their breasts removed or enrolling in the ongoing trial of tamoxifen. Scientists hope the drug will help prevent many breast cancers but its effectiveness is unknown and it has many potential health risks.

Dr. King's lab is also doing two projects on AIDS research, asking whether genetic variations could explain why some people survive with the disease much longer than others. Her team is studying the genetics of systemic lupus, an autoimmune disease in which the skin and joints are progressively destroyed, and it is hunting for the gene behind hereditary deafness.

Dr. King is an ardent proponent of the Human Genome Diversity Project, led by Dr. Luca Cavalli-Sforza, a population geneticist at Stanford University, in which researchers plan to sample genetic material from some 400 human populations worldwide, with an emphasis on the oldest and least intermixed peoples, like the Basques of Spain, and the Ket and Gilyak of Siberia. By scrutinizing the chemical runes of genes, the researchers hope to answer many questions of evolutionary, linguistic and anthropological sweep: where did modern humans come from? How did they migrate across the globe? Did genetic changes in any way correlate with language variations, and might genetic discrepancies explain differing rates of disease among different countries? Dr. King spends several days each month traveling to Washington, in part to lobby for money to support the enormously complex effort.

This harlequin collection of projects is done by a relatively small lab of 20 people, including Dr. King. She also teaches graduate and undergraduate courses, including a freshman genetics class for nonscience majors. And, rare for a research professor, she sees teaching not as drudgery but a pleasure. Yet in spite of it all, she manages to look young for her age, with a slab of dark hair that seems almost immobilized by its own thickness. Doesn't she ever feel overwhelmed?

"Of course I feel overwhelmed!" she said, her voice rising up in a whoop ever so slightly reminiscent of Julia Child. "What does being overwhelmed have to do with it?" She walks and talks so swiftly that by comparison one feels trapped in resin.

"She's insightful, irreverent, energetic, a wonderful antidote to the notion people have of scientists as lifeless, bloodless drones," said Dr. Eric Lander, a molecular biologist at the Whitehead Institute for Biomedical Research, who is an old friend.

Dr. King traces her scientific style to her mentor and thesis adviser at Berkeley, Dr. Allan C. Wilson, an intellectual firebrand who died in 1991 of cancer at the age of 57. Dr. Wilson was famed for his work on the so-called genetic Eve, a woman who supposedly lived about 100,000 years ago in Africa and is the theoretical mother of all humans alive today. Those who worked in Dr. Wilson's lab mastered the art of attacking evolutionary puzzles with molecular artillery, relying particularly on the genes sequestered in the mitochondria, the tiny powerhouses of the cell.

Dr. Wilson also stopped Dr. King from quitting science almost before she got started. "I could never get any of my projects to work, and I was very depressed and distracted," she said. "He said, if everybody who couldn't get anything to work dropped out of science, there would be no science." Thus inspired, she completed her Ph.D., showing, to the shock of herself and the entire scientific community, that humans and chimpanzees have more than 99 percent of their DNA in common.

From there she went to Chile with her husband, Robert Colwell, a zoologist, to teach, but they decided to return to the United States after the leftist government of Salvador Allende was overthrown. Thanks to her experience in South America, her familiarity with the language and people, when the Argentinian grandmothers sought the help of scientists to solve the problem of the missing children, Dr. King ended up as the molecular geneticist on the case. The work was at once grueling and inspiring, demanding frequent trips to Argentina, 18-hour days and the spine to stand up against the surly and grudging military there. The Argentinian project continues, and so far 53 children have been reunited with their original families, with another 150 yet to be found. They would now be teenagers, and could be anywhere: in Latin America, across Europe, in the United States.

Throughout the grandmothers project, Dr. King had the added spur of knowing that the kidnapped children were the age of her own daughter, Emily Colwell. Dr. King's marriage broke up when Emily was 5, leaving Dr. King as a young, single mother struggling to succeed in a field known for its lengthy, irregular hours and its relentless pace. She believes that one reason her marriage failed is that, as a scientist, "you can only do two out of three," adding: "I was a young mother, a young scientist and a young wife. Something had to collapse, and it was the marriage." Now her daughter is a senior in high school who plans to go into constitutional law.

In her own lab, Dr. King tries as much as possible to accommodate those with families. True to the capacious spirit of Berkeley, she has always had many female graduate students and postdoctoral fellows in her lab, as well as minorities and homosexuals. "I was in this business for years and years before I had a straight white male graduate student," she said.

Yet there is a limit to how freewheeling any high-powered lab can be, and many of her students practically live there, particularly those who are involved in the breast cancer project, where outside competition is keenest. Dr. King says she hates competition, and she sees it as one way in which men have put a needlessly and tediously masculine stamp on the profession. Once she showed up at a meeting with a pencil case in the shape of a shark, given to her by the child of a friend.

"She put the pencil case on the table and said, 'This is for all the sharks in the audience,'" said Dr. Anne M. Bowcock of the University of Texas Southwestern Medical Center in Dallas, one of Dr. King's closest friends. "Nobody laughed."

But as the men who know her are quick to point out, Dr. King is no wilting lily. She speaks her mind, goes after what she wants and will never cede her ground.

When it comes to competition, said Dr. Raymond L. White, a friend and competitor at the University of Utah in Salt Lake City, "Mary-Claire is at no disadvantage, no disadvantage at all."

—NATALIE ANGIER
April 1993

THE PERILS OF GENETIC KNOWLEDGE

Dr. Mary-Claire King may have wanted to clone the first gene for hereditary breast cancer very, very badly—as a colleague of hers said, "It was her reason for getting up in the morning"—but she wasn't a bad sport when she lost the overheated race in 1994. Instead, she praised Dr. Mark Skolnick of the University of Utah and his colleagues for their "lovely" and "beautiful" work—and immediately set out to put the discovery of the gene, called BRCA1, through its paces. When researchers announced the isolation of another gene linked to breast cancer, called BRCA2, soon afterwards, that gene was added to the mix of biomysteries to be solved.

Dr. King and others sought to determine what the findings mean to women everywhere, who should be tested for mutations in the genes and what a woman should do if she finds she is a carrier of an inherited mutation and is thus at greatly heightened risk of contracting breast and ovarian cancer alike.

Today, researchers have some sense of the worthiness of genetic screening for breast cancer. Women who come from families with a notable history of breast or ovarian cancer—particularly when the diseases struck before the women reached menopause—are advised to consider being tested. In addition, Dr. King and others have learned that women of Ashkenazi Jewish heritage are much more likely than the general population to carry one of three specific mutations in the breast cancer genes. The scientists believe that the three mutations are ancient flaws that have been passed along through the ages because Ashkenazi Jews often marry others of their group, and then have their children before any hereditary cancer has made its unwelcome appearance.

Thus, Dr. King suggests that a Jewish woman who has even a modest amount of breast cancer in her family might opt to be screened for the three so-called Ashkenazi mutations.

As for what a woman can do if she learns she is a carrier, Dr. King admits that the current alternatives are sorry and harrowing ones. "Women who have found themselves to be at extremely high risk have often chosen to have prophylactic surgery," she said. "They've had their breasts and ovaries removed before the organs could turn cancerous.

"We've saved lives with genetic screening, there's no question about it, but it's been at the cost of oophorectomies and mastectomies. There's got to be a better solution to this."

Dr. King and many others are struggling to apply their knowledge of the breast cancer genes to the design of improved therapies, but so far to scant avail. "I think biologists and geneticists are on the right track, but it's a very difficult problem," she admitted. "Meanwhile, you want to shake yourself because women keep dying."

—N.A.

David A. Summers

Out of the Mines and Into the Lab

Bill Stover *for* The New York Times

Deep underground, surrounded by the constant coolness of rock and the sound of seeping water, miners cleave the Earth with a powerful tool. Above ground in metal-fabricating factories, carpet-making companies and food-processing plants around the nation, similar tools carve away carefully and precisely, without benefit of a blade.

Deceptively simple in principle, this tool, the high-powered water jet, together with its relatives using other fluids, has come to play a role in countless aspects of modern technology. And one of its leading champions, Dr. David A. Summers, a 50-year-old mining engineer from England, is constantly searching for new applications in mining and elsewhere.

331

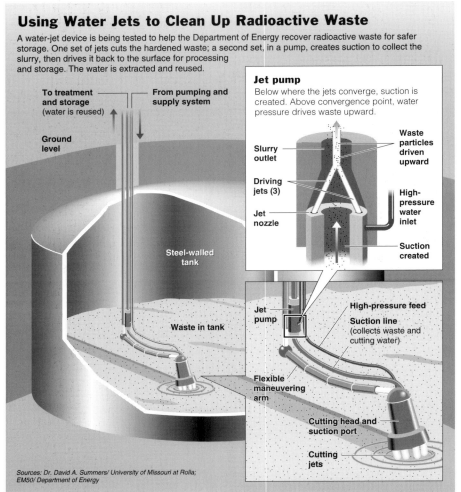

Using Water Jets to Clean Up Radioactive Waste

A water-jet device is being tested to help the Department of Energy recover radioactive waste for safer storage. One set of jets cuts the hardened waste; a second set, in a pump, creates suction to collect the slurry, then drives it back to the surface for processing and storage. The water is extracted and reused.

To treatment and storage (water is reused)

From pumping and supply system

Ground level

Jet pump
Below where the jets converge, suction is created. Above convergence point, water pressure drives waste upward.

Slurry outlet

Waste particles driven upward

Driving jets (3)

Jet nozzle

High-pressure water inlet

Suction created

Steel-walled tank

Waste in tank

Jet pump

High-pressure feed

Suction line (collects waste and cutting water)

Flexible maneuvering arm

Cutting head and suction port

Cutting jets

Sources: Dr. David A. Summers/ University of Missouri at Rolla; EM50/ Department of Energy

The New York Times; illustration by Frank O'Connell

"Because of environmental concerns, water jets are the future of mining and digging because they can do the job better and cleaner than before," he said in a visit at the University of Missouri campus in Rolla, where he directs the High Pressure Waterjet Laboratory. "But even beyond this, the water jet has so much potential to be useful at other things that we are constantly amazed."

While anyone who has used a garden hose to clean a sidewalk knows that forcing a thin stream of water through a nozzle can make a useful tool, the technology has become increasingly sophisticated in the hands of engineers like Dr. Summers.

In the last two decades, the traditional use of water jets as cleaning tools has expanded to include removing old concrete from buildings and other structures before restoration, cleaning airport runways of old tire rubber and grit, paint removal and asbestos abatement.

High-pressure water jets have become the mainstay of so-called cold cutting, replacing or complementing blades and lasers in slicing everything from high-strength metals to paper. Water jets avoid the heat damage of lasers and the edge-crushing effects of blades while leaving little dust and debris.

Most of the carpet and large plastic components, like dashboards, that are used in new cars are now cut by water jet. Much of the cutting needed to make disposable diapers, paper towels, toilet paper and corrugated boxes is done by water jet. Lengths of these materials can be moved so fast under the thin stream of water that there is virtually no wetting, experts say.

In addition, jets of water or oil are used extensively in the food industry, cutting frozen pizza and the familiar uniform bricks of frozen vegetables, as well as chicken nuggets, candy bars and baked goods.

Dr. Summers got a gritty start on his path to the academic life and eventual development of his specialty. He was born in Ashington, a coal-mining community near Newcastle in northeastern England. At the age of 17, he followed a family tradition and became the eighth generation to take up mining. He worked in the mines for 15 months, helping to build lumber supports, loading coal by hand and working the wall with a shovel, pick and ax. This stint as an "indentured apprentice" for $10 a week "was a strong incentive to go to the university," he said.

At the University of Leeds, where he earned his undergraduate and doctoral degrees in mining engineering, he was introduced to water jets when a faculty member who had started a research project left and he took it over. After receiving his Ph.D. in 1968, he came to Missouri to teach and brought his water jets with him.

Married for 22 years to the former Barbara Muchnick, who teaches Spanish and piano, Dr. Summers is the father of two sons, one in high school and the other in medical school. His interest in computers recently led him to discover computer animation, which he is using to create presentations to explain water jet technology and proposals for new projects.

The walls of Dr. Summers' spacious office in an old brick building that once housed a hospital are lined with floor-to-ceiling bookshelves brimming with books, papers and artifacts on subjects ranging from energy policy to medieval history. On the top shelves, circling the room, are scores of figurines and statues depicting miners, as well as numerous prints and pictures about mining. He proudly points to a framed certificate from the National Mining Board in England, dated September 21, 1961, accrediting him as an apprentice miner.

At the Missouri laboratory, faculty members and graduate students develop new ways of using fluid-jet technology for mining, industrial cleaning and cutting. For instance, the laboratory is one of several working on ways of using cutting jets to destroy military rockets and artillery shells and recycle their components. In a test mine at the school, workers under a Navy contract have shown that they can disassemble surplus shells with water jets and convert the contents into commercial-grade explosives.

Researchers are also designing water-jet equipment for the Department of Energy to remove radioactive waste from leaky containers so it can be treated and stored more safely. Dr. Summers and his colleagues are testing a multijet, rotating cutting and extraction assembly that recovers the little water it uses to avoid adding much to the volume of waste removed.

"We see using water jets to handle disposal of toxic or dangerous substances as a growth area," Dr. Summers said.

The technology also has promise as an excavation tool in difficult locations, as Dr. Summers and a team of staff members and students proved in 1991 when they used water jets to carve space for a 350-seat underground theater near the Gateway Arch in St. Louis. The team excavated 2,000 cubic yards of limestone to create a cavern within 50 feet of the southern leg of the arch.

The Missouri researchers also took water jets into another unconventional area this year when they began developing a new method for treating the most common type of skin cancer, basal cell carcinoma. Teaming up with Dr. William Van Stoecker, a Rolla dermatologist, Dr. Summers and his colleagues began experiments with a water-jet device that they believe can remove skin tumors without damaging small blood vessels below.

This method, using a fan-shaped jet nozzle to deliver a sterile saline spray, slices away bits of diseased tissue without greatly disturbing healthy skin, which should lead to less bleeding, more rapid healing and less scarring than conventional treatments, they say.

A decade ago the water-jet enthusiasts sought to demonstrate the technology by creating on campus a half-scale partial reproduction of Stonehenge, the rings of stone columns erected many centuries ago on Salisbury Plain in England.

The water-jet process depends upon factors like the pressure applied to the liquid, the size and shape of the nozzle holes, the rate of flow, the temperature of the fluid and the addition of agents like abrasives. Water jets in a car wash typically fire two to three gallons of water each minute at pressures of 1,500 pounds per square inch, while those used for industrial metal machining can fire a stream of hair-thin, abrasive-laced water at less than a gallon a minute at a pressure of more than 50,000 pounds per square inch.

After almost three decades of research, Dr. Summers has written *Waterjetting Technology*, a book about the history, current techniques and future of this specialty, with E. & F. N. Spon of Britain as publisher.

Dr. Andrew F. Conn, an independent water-jet consultant in Baltimore who has been a friend and competitor of Dr. Summers for 20 years, said, "There are a million potential applications out there for water jets." He praised Dr. Summers for his role in popularizing the methodology. "David is a true messenger of the technology," Dr. Conn said. "He has the time and energy to go worldwide and spread the word. His use of self-deprecating humor brings people in and, when he's captured the audience, he sells them on water jets."

Dr. George Savanick, president of the Water Jet Technology Association, a professional group based in St. Louis, said no one knew exactly how big the industry was, but he estimated that its goods and services were worth billions of dollars each year. Dr. Mohamed Hashish, vice president for water jet technology at Quest Integrated Inc., a private research and development company in Kent, Washington, said the industry was set for enormous growth as new technology became more common. Dr. Hashish, who pioneered abrasive jet cutting 15 years ago, said water jets would be used more in advanced machine tooling, using computer control to have them mill and turn precision parts, for example.

In the 1970's, Dr. Summers said, water-jet mining machines essentially solved the dust problems of coal mining, but little money is now going into mining research. "We need energy, and oil supplies are declining," he said. "Nuclear is going nowhere, and the prospects of renewable resources are uncertain. We'll need energy from the ground for a while, but we're in danger of losing our mining expertise in a few years."

Growing animated in a discussion of his two primary interests, water jets and mining, Dr. Summers swept back his graying hair and continued in earnest: "The water jet eliminated the problem of black lung. Tied to robots, it could be the future of safe mining, a tool to selectively mine and reduce the environmental impact of mining."

Whatever lies ahead for water-jet technology, it has left its mark on the campus with the Stonehenge reproduction. It was executed with backing from alumni and a chancellor who was an archeoastronomer and was based on the design of Dr. Joseph Senne, an astronomer and civil engineer.

Using a water-jet lance with a spinning nozzle that fired two streams of water, the Summers team cut through 160 tons of granite a half inch at a time to construct slabs for a circular arrangement of five 13-foot-high stone structures, each consisting of two upright stones capped by a horizontal one. The structures, positioned so that the Sun and Moon are aligned with them at the time of the solstices, were arranged within a circle 50 feet in diameter that was formed by 29 rectangular stones, each one foot high.

Dr. Summers, proudly explaining construction of the megalith, noted that experts estimated that the monument would decay about one inch every thousand years. "It should be here for a long time," he said.

—WARREN E. LEARY
January 1995

OH, THOSE CREATIVE ENGINEERS

Engineers tend to be the Rodney Dangerfields of science. Too often, those in the profession say, scientists or designers come up with ideas and turn them over to engineers to implement, without appreciating the ingenuity it often takes to turn an idea into something that works.

"Engineers don't get the credit and, after a while, often don't expect it," Dr. David Summers says. "It's a fact of life you learn to live with."

But turning an idea or concept into something that is practical and economical is an essential part of the creative process, he says. If engineering is to attract the brightest students, the profession has to showcase the imaginative aspect of the work and stimulate interest among prospects before they enter college, he says.

Dr. Summers's latest work with water jets could be a case in point. Recently elected president of the International Waterjet Society, Dr. Summers has been pushing the boundaries of this technology. In his laboratory and elsewhere, water jets have moved from being tools for dust-free mining and industrial cleaning into such new areas as medicine, toxic waste disposal and military applications.

Working with dermatologists, Dr. Summers and colleagues developed and licensed a new method for removing basal cell skin cancers with a sterile water-jet spray that minimizes damage to underlying tissue. And equipment developed for the Department of Energy to remove radioactive waste from leaky containers has successfully been used at the Oak Ridge National Laboratory in Tennessee.

With support from the Pentagon, Dr. Summers's High Pressure Waterjet Laboratory recently developed a new approach to detecting and removing land mines. Hundreds of millions of forgotten metal and plastic land mines kill and wound thousands of people yearly. Finding simple and inexpensive ways of detecting and destroying them has become a priority for the United Nations and humanitarian groups worldwide.

The approach coming from Dr. Summers' laboratory starts with a simple principle that many have noticed but that engineers took a step further. "When you squirt a water pistol, the sound of each target it hits is different," he says. "The splash sounds different on paper, wood, glass or anything else you hit. It turns out that when you send a water jet into the ground, you get an acoustic signal back that's different depending on what it hits underneath."

By mounting on a remote-controlled cart an array of jets that fire short bursts into the ground as it moves, the reflected sound tells whether objects detected below are plastic, metal, rock, wood or other materials. To investigate a suspicious item, water jets attached to a pumping device liquefy and remove what is covering it so it can be identified. "If it's a land mine, we can turn on high-pressure water jets that cut up the mine and fuses without setting it off," he says.

Dr. Summers says most of his new projects are coming from industrial contracts that he can't discuss for proprietary reasons. But he notes with pride that interest in water-jet technology is growing. "This whole business is gradually sneaking its way into being a large industry," he says. "And I couldn't be more pleased."

—W.E.L.

Ellen J. Langer

A Scholar of the Absent Mind

Rick Friedman *for* The New York Times

Dr. Ellen J. Langer's specialty may seem a little odd for a psychologist: she studies mindlessness.

Everyone exhibits it, of course. People misplace their keys. They enter a room only to realize they don't know why. They talk to mannequins before realizing that a reply isn't likely.

One of Dr. Langer's favorite examples of mindlessness concerns the time she used a brand-new credit card in a department store. The clerk noticed she had not yet signed it, and handed it to her to sign the back. After passing the credit card through the machine, the clerk handed her the credit card receipt to sign.

"Then she held up the receipt I signed next to the card I had just signed, and she compared the signatures!" Dr. Langer said. "Amazing!"

Dr. Langer's studies have always centered on the degree to which humans are in control of their actions, or rather, the degree to which they maintain the illusion of control. Beginning in the late 1970's, her studies showed that mindless behavior was far more widespread than people liked to believe.

Based on that and other work, she became the first woman to be named a tenured professor in psychology at Harvard University. In her recent work she has followed up her studies of mindlessness by working on antidotes—the "mindful" attitude and how to cultivate it. She has written two popular books published by Addison-Wesley, *Mindfulness* (1989) and *The Power of Mindful Learning* (1997).

Her latest book argues that traditional methods of learning can produce mindless behavior because they tend to get people to "overlearn" a fact or a task and suggest that there is only one way to do it. She argues that it is important to teach skills and facts conditionally, setting the stage for doubt and an awareness that different situations may call for different approaches or answers.

Dr. Langer, 50, was born in the Bronx, grew up in Yonkers, and earned a bachelor's degree in psychology at New York University and a Ph.D. in psychology at Yale University. Colleagues say she is about the least mindless person they know. She is, as one colleague put it, "aggressively thoughtful" and full of creative energy.

Dr. Langer, said Dr. Robert Abelson, professor of psychology at Yale, "enjoys being outrageous, and challenging conventional wisdom," adding, "Dr. Langer even as a student noticed that while psychologists were always talking about thinking and behavior, it seemed to her that people were behaving thoughtlessly just as often."

Psychologists have been aware for at least a century that some complicated behavior is performed automatically, that is, without conscious deliberation. As Dr. Langer wrote, citing people who say hello to a mannequin or write a check in January with the last year's date, "when in this mode we take in and use limited signals from the world around us (the female form, the familiar face of the check) without letting other signals (the motionless pose, a calendar) penetrate as well."

People have in their repertoires thousands of "scripts" for talk or behavior that they act out when they are cued by something familiar. The array of behavior people can carry out without thinking is enormous.

When Dr. Langer started her work it came as something of a challenge to mainstream psychology. When her career in the shortfalls of thinking began in the 1970's, psychology as a whole seemed to be moving in the opposite direction, turning to cognitive psychology in a big way. A rush of studies had begun on the details of human thinking and the importance of reasoning in human behavior.

"In the midst of that," said Dr. Daniel Wegner, a research psychologist at the University of Virginia, "Ellen Langer started saying, 'Maybe, but we really are pretty mindless.' In the history of psychology there had been some work on mindlessness, or automaticity, but she revived it and extended it much farther."

Dr. Wegner said that Dr. Langer's work was important and had "really broadened our perspective in psychology," adding: "She showed that mindless behavior is fairly widespread and general. She showed us that we have to take into account not only things that make sense, but the things that don't make sense. She has helped make us rethink the role of thought in behavior."

In the 1970's and 1980's, Dr. Langer carried out a series of landmark studies to make the point scientifically, the most famous of them referred to as "The Copy Machine."

In that study, she stationed someone at a copy machine in a busy graduate school office. When someone stepped up and began copying, Dr. Langer's plant would come up to the person and interrupt, asking to butt in and make copies. The interruption was allowed fairly often, about 60 percent of the time. But the permission was granted almost 95 percent of the time if the person stepping up to interrupt not only asked, "May I use the copy machine?" but added a reason, "because I'm in a rush."

That seems to make sense. People heard the reason and decided they were willing to step aside for a moment. What was odd, Dr. Langer found, was that if the interrupter asked, "Can I use the machine?" and added a meaningless phrase, "because I have to make copies," the people at the machine also stepped aside nearly 95 percent of the time.

The idea, she said, is that the listener at the copy machine heard a two-part statement: a request and something like a reason. That was all their mental script for such a situation required. They never did reflect on the fact that the interrupter's "reason" was not meaningful.

The people at the copy machine only began to listen more carefully to the request and judge the reason for it when those interrupting had a large number of pages to copy. They were asking for a big favor, and in this case, adding "because I have to make copies" had no effect. Only when people added that they had to hurry would people step aside.

Dr. Langer's most influential studies were some of the earliest to establish the notion of "learned helplessness" in nursing homes. She and a co-author, Dr. Judith Rodin, now president of the University of Pennsylvania, found that if the elderly residents were given some control in their lives—such as when to watch movies, where to visit with relatives, whether and how to raise plants—they fared better. They felt better, and were more able to handle visitors and tasks in the home. In follow-up studies over the next two years the investigators found those in the group given some control continued to be more active, and in fact, ultimately lived longer.

All this apparently stemmed, Dr. Langer said, from an intervention by experimenters that lasted only a few weeks. "That so weak a manipulation had any effect suggests how important increased control is for these people, for whom decision-making has been virtually nonexistent," she wrote in a paper.

In one of the simplest experiments, she asked elderly patients to put together jigsaw puzzles, and measured their success. One group was simply asked to do it, another given "encouragement only" and a third was actively helped by the nursing staff to assemble the puzzles. She found that those who were helped performed most poorly and considered the task difficult, while those who did the work themselves performed better and found the task easier.

"When you provide people with opportunities to make choices, they live longer, they are happier," she said in an interview. "Sometimes the situation in nursing homes goes to the opposite extreme, and people entering a nursing home essentially give up their personhood. The staff ends up talking to visiting relatives about care of the elderly and ignoring the elderly themselves."

After spending the first part of her career demonstrating that people often behave mindlessly, Dr. Langer is now showing how people can overcome that tendency. She says she also vaguely recalls switching from the negative to the positive after someone suggested that in psychology people studied their own worries. "I didn't want to be thought of as mindless," she jokes, "so I changed the name of what I was studying from mindlessness to mindfulness."

Dr. Langer, an avid tennis player, uses an example from that sport: "At tennis camp I was taught exactly how to hold my racket and toss the ball when serving. We were all taught the same way. When I later watched the

U.S. Open, I noticed that none of the top players served the way I was taught, and, more important, each of them served slightly differently."

Because each person has different attributes—hand size, height, muscle development—one way of serving cannot be right.

Thus it is important to teach everything conditionally. "For example," she said, "you can teach 'here is one way of serving,' or, 'if an incoming ball has backspin, here is one way that you can use to return it.'" This method accounts for the fact that each instance is different, and a person's responses must be changed moment to moment, day to day.

In her studies of learning, she found that just changing the way instructions are worded, from "this is the answer" to "this is one answer" can have a significant effect on how facts or tasks are learned.

In one study, high school students were taught lessons in physics by videotape. But with half the students, the video was preceded by verbal instructions that the tape was only one way of looking at the physics problems, and that students should consider other ideas and methods. When tested, both groups were equally able to repeat the facts given in the videotape, but only the group with the extra instructions tended to use other methods and information from their previous experience.

In another study, students who were starting piano lessons were divided into two groups, those who would learn the scales in the traditional rote manner and those who would learn them "mindfully." The chief difference was that the "mindful" group was instructed to be creative and to vary their playing as much as possible.

When independent evaluators later listened to tapes of the students, without knowing which were assigned to each group, they rated the "mindful learners" as more competent overall and more creative, Dr. Langer wrote.

Other studies show that memory is improved by "mindful" learning, even down to the simplest task of trying to remember a series of pictures.

Rote learning can be destructive not because repetition is unnecessary—tennis players may hit thousands of similar forehand shots before attaining some competence—but because it teaches a person to use one mindless response. "Teaching skills and facts in a conditional way sets the stage for doubt and an awareness of how different situations may call for subtle differences in what we bring to them," Dr. Langer wrote.

She has done studies showing that when people learn by rote, the small steps that make up the skill come together in larger and larger units. The smaller components may thus be "lost," so people are then unable to vary them. Yet it is by adjusting and varying these pieces that people can improve their performances.

It is vital to include the possibility of different answers at the time of learning, because once a rigid pattern of response is set, "it is hellishly difficult to change," Dr. Langer said, adding, "But if you learn some flexibility from the beginning, it's much easier to vary your response."

—PHILIP J. HILTS
September 1997

DEBUNKING THE RATIONAL DECISION

Dr. Ellen Langer continues to work on her notion of mindlessness versus mindfulness, and to apply it to different areas of life and psychology. Lately, she has been venturing into an area called decision theory.

Throughout the history of psychology there has been debate about how people make decisions and to what degree decisions are rationally made, consciously weighing costs and benefits, and to what degree choices are guided by irrational feelings and impulses.

Dr. Langer suggests that decisions are neither rational nor irrational, but arational. To make a decision, she said, most people gather information about the choice they are to make and eventually reach a point at which they simply make a "cognitive commitment"—a mix of unconscious feelings and attitudes, combined with a few conscious thoughts about what they like and don't like. In this process, there is no actual list of pros and cons being made and consciously evaluated, as psychologists in the past have suggested, but simply an accumulation of bits of feeling and information that finally tip the scale one way. On reflection, people finally "realize" they have come to a conclusion. Only after arriving at a conclusion, she said, do people usually begin to analyze the decision's implications or to tote up pluses and minuses about it.

Given this normal pattern, she advises others to give in and realize they cannot really make a rational decision based on lists and cost-benefit weights. In fact, such analyses tend to ignore important parts of thinking like emotions and tend to be limited to conventional categories.

Rather, she says, the most important part of decision making is not the conscious analysis, but the gathering of information and trying to see a fuller range of choices.

For example, on multiple-choice questionnaires, little real decision making is going on. Dr. Langer calls this passive deciding.

The better option, when available, is active deciding, in which the first step is to generate options by examining one's own preferences and by creating many choices beyond a few obvious possibilities.

In one experiment, she gave one set of subjects a case in which they had to choose how to resolve a conflict between themselves and another party. They were asked whether they would give in, stand firm, or agree to a named compromise. "But when we gave people conflict situations to respond to with an explicit instruction to resolve the conflict without exercising any of these options, thus provoking active deciding, people often found a creative, win/win solution," she said. "Subjects not so instructed selected the compromise solution where no one's needs were fully met."

"Since psychologists have always focused on trying to determine what the preferences of the deciders are and how they should match them to the choices given, we hold the world constant and fail to realize all the information we might reasonably have considered if we had let it vary more naturally, or if we actively sought to broaden the number of choices available," she said.

It is harder to make decisions when real thought is required, she says. But on the other hand, people feel more in control if they make what she calls mindful choices, and the worry about what is a good or a bad decision seems less important.

—P.J.H.

Günter Wächtershäuser

Amateur Shakes Up Ideas on Recipe for Life

Horst A. Friedrichs/Agency Anne Hamann

Scientific research is so specialized, and its hot fields so crowded, that it is highly unusual for an amateur to make a significant contribution. So it is surprising that a patent lawyer from Munich, Germany, has managed to elbow his way to the forefront of attention with a striking theory about the origin of life.

His name is Dr. Günter Wächtershäuser (pronounced VEK-terz-hoizer), and his success is the more remarkable because instead of starting with facts, a customary point of scientific departure, he has erected a grand theory about how life began, from which he makes and tests deductions. Proof of one element of the theory was recently published in the journal *Science*.

His ideas, if correct, mean that scientists seeking to understand the origin of life on Earth should focus on the chemistry that takes place on the surface of minerals, particularly in hot environments like the deep undersea volcanoes that gash the ocean floor. Until now, lightning strikes and tidal pools have been leading suspects for pre-biotic chemistry, the chemical reactions that presumably preceded the emergence of life.

Several scientists who are familiar with Dr. Wächtershäuser and his work hold him in high regard. "He is a great mind and I don't say that lightly," said Dr. Carl R. Woese, a microbiologist at the University of Illinois. Dr. Christian de Duve, a biochemist at Rockefeller University with an interest in the origin of life, describes him as "a very special kind of character, very likable," but notes his idiosyncratic way of doing science.

Dr. Wächtershäuser is no innocent in scientific matters. He has a doctorate in organic chemistry from the University of Marburg. But according to Dr. Woese, who sparked his interest in the origin of life, "he became disenchanted with the German system," adding, "He determined chemistry wasn't for him and became a lawyer."

The other major influence on his work was the late Sir Karl Popper, a philosopher who argued that true sciences proceed by setting up a theory and trying to prove it false. Following the Popper prescription, Dr. Wächtershäuser decided the right way to investigate the origin of life was first to construct a theory about how it must have originated.

This bold step was in contrast to the usual approach. Whether because of the sheer difficulty of trying to imagine the natural chemistry of the distant past, or from trepidation at presuming to decide how life began, most scientists have followed a trial-and-error method in studying the likely chemical predecessors of living cells. They choose ingredients deemed to have been present in the primitive Earth's ocean or atmosphere and cook them up to see if any promising chemistry occurs.

"I used to believe in the soup theory," Dr. Wächtershäuser said, referring to the ingredient-cooking experiments, "but when I met Carl Woese I found the soup theory was not sound. So I thought casually about an alternative system. Then, since we were friends of Karl Popper, we visited him in England. During breakfast I mentioned my ideas on an alternative and he urged me to work it out."

Dr. Wächtershäuser's theory has caught the interest of scientists who have come to believe for other reasons that geothermal vents are a likely site of life's creation. His argument goes as follows:

The soup theory, to use his term, assumes that the chemicals that preceded life came together in three dimensions. But chemicals moving freely in air or water do not stay assembled for long. On a surface, however, they are more stable, so in his view the first chemical reactions must have been on a two-dimensional surface.

"It's like looking at airplanes in the air and asking, 'How do they get made up there?' " Dr. Wächtershäuser said. "Of course, they get made on the ground. The origin of life is on the surface, and only the best organisms took off and conquered the third dimension."

The surface on which the prebiotic chemicals formed, he inferred, must have been open to water but not enclosed—an important point because many ideas about the origin of life assume that the first chemical reactions started in bubbles or other enclosures. The surface must also have been positively charged so as to bind the chemicals typical of life, many of which carry a negative charge. Free to move around on the surface, the chemicals could try out different mixes and matches until they hit upon the route to prebiotic reactions.

This line of thought, Dr. Wächtershäuser said, led him quickly to consider minerals as a likely surface, and specifically to focus on metal sulfides as a promising two-dimensional cradle of life.

Ideas about the origin of life, he noted, fall into three classes, depending on which critical element of life is deemed to have started first. In some schemes, the cell membrane is thought to have arisen first. The problem with membrane-first theories lies in explaining how food chemicals get into the cell. Another idea is that the nucleic acids came first. But even the simplest nucleic acids are complex molecules.

The third kind of approach, that of Dr. Wächtershäuser, is that life begins with a metabolism—a repeated cycle of chemical changes—and that the metabolism invents the cell membrane, the genetic machinery and everything else.

The metabolism would have to have been a carbon fixation cycle, one that joins two carbon atoms together at every cycle.

Chemicals like amino acids, the building blocks of proteins, would have formed first as useless by-products of the metabolism, then would have become catalysts, or agents of chemical change, that directed more of their own synthesis.

Nucleic acids, in Dr. Wächtershäuser's view, also arose as by-products and then became catalysts of their own synthesis. Later, like the queen bee in a hive, they acquired responsibility for the whole system's reproduction.

With all this chemistry going on at the metal sulfide surface, sooner or later some bundles of chemicals managed to cloak themselves in a membrane and escape from their two-dimensional world. That event, according to the theory, marked the birth of the first cells.

Just as few military strategies survive contact with the enemy, even the most beautiful scientific ideas have a distressing tendency to collapse on their first encounter with experiment. But Dr. Wächtershäuser recently obtained experimental support for a significant element of his theory: the carbon fixation cycle.

Bacteria possess an ancient metabolic pathway that synthesizes acetic acid, a simple chemical based on two carbon atoms stuck together. The acetic acid, because it is coupled to another compound, is in an active form that is capable of reacting with other chemicals. Seeking to reproduce a prebiotic version of this reaction, Dr. Wächtershäuser took gases of the kind exhaled by deep sea vents and mixed them in the presence of iron and nickel sulfides.

He found that an active form of acetic acid was produced, which he regards as a likely candidate for the metabolism that initiated life. This is the result that was published in *Science* with a co-author, Claudia Huber of the Munich Technical University.

The soup theory experiments, Dr. Wächtershäuser says, "produced many wonderful compounds but they were all inactive," meaning that they could not react together to produce the more complex chemicals of life. His experiment produced acetic acid in activated form, "a complete analogy to what happens in biochemistry, and it runs without any enzymes," Dr. Wächtershäuser said, referring to the natural catalysts of living cells.

Dr. Stanley L. Miller of the University of California at San Diego, who might be regarded as the chief chef of the soup theory, has vigorous reservations about Dr. Wächtershäuser's theory. It was Dr. Miller's famous experiment of 1953 that first showed how amino acids could be generated from the chemicals present in the early atmosphere and ocean.

His own experiments in prebiotic chemistry, Dr Miller noted, produce significant amounts of life-suggestive chemicals like amino acids, none of which have been made by Dr. Wächtershäuser's system.

"The theory as a whole is this overblown thing which so far has not been shown to work," Dr. Miller said. "Making acetic acid is sort of blah to me, that's nothing." He doubts that surfaces really make chemicals more stable, and as for life originating in geothermal surroundings, he said, "the high-temperature origin of life is out of the question."

Some scientists, like Dr. de Duve of Rockefeller University, believe there are weaknesses in Dr. Wächtershäuser's overall theory but find many of his ideas of interest. Others, like Dr. Norman R. Pace of the University of California at Berkeley, believe Dr. Wächtershäuser has created a significant alternative to the approach of Dr. Miller and others. His work has touched off a small "paradigm shift, turning from solution chemistry to surface chemistry," Dr. Pace said.

Dr. Wächtershäuser said he had delayed publishing his theory for fear of ridicule. At first he was not criticized. "Later on I met the wolves, but then I had grown to a certain size," he said, having now had articles published in several scientific journals.

Asked how he would feel if his theory were indeed correct and he succeeded in creating life in one of his experiments, Dr. Wächtershäuser replied that that would be a matter of definition. A cycle of chemical reactions on a surface "cannot die," he said. "One can say that what cannot die isn't life but one doesn't know. I don't believe in defining life. I am not interested in definitions, only in the mechanism."

—Nicholas Wade
April 1997

OTHER IDEAS ABOUT THE BEGINNING

There may be no more difficult problem in science than explaining the origin of life. The reason is that life originated on Earth apparently only once and under physical and atmospheric conditions that no longer exist. How life began is thus in large part a historical question that cannot be settled by scientific experiment.

Still, scientists have tried to reconstruct plausible theories for how life might have begun. During the 20th century there were three major conceptual advances, to which Dr. Günter Wächtershäuser, a patent attorney in Munich, Germany, may possibly have contributed a fourth.

The first was the suggestion in 1924 by Dr. V. I. Oparin and other scientists in the Soviet Union that the first living cells may have arisen during the natural chemical processes that occurred on the early Earth. Dr. Oparin's theory was devised at the request of the Communist Party of the Soviet Union, which wanted a scientific theory of the origin of life to bolster its atheistic campaign. Despite these unpromising auspices, Dr. Oparin's thesis seemed a step in the right direction, even though it was vague as to the exact mechanism by which life might have arisen from his postulated soup of pre-biotic chemicals.

These details were provided, at least in part, by a famous experiment in 1953 in which Stanley Miller, then a graduate student, brewed up a mixture of water and gases designed to mimic the early Earth's atmosphere, along with frequent electrical discharges to simulate lightning strikes. In the tar at the bottom of his flask Mr. Miller found that amino acids and other important constituents of living cells had been formed. The finding seemed to be a brilliant first step in showing how to get from Oparin's pre-biotic soup to the first living cells.

But in the years since 1953, no one has managed to take Miller-type experiments much further. The early Earth's natural ingredients will make a tar-like sludge, but a sludge is a long way from a living cell.

The third significant step in origin-of-life studies came from biologists working on RNA, a nucleic acid very similar to DNA. Dr. Sidney Altman and Dr. Thomas Cech showed RNA could act as a chemical catalyst or enzyme as well as store information. Their discovery seemed to resolve the chicken-and-egg problem of whether proteins or DNA came first in the evolution of life by showing that an RNA molecule could in principle perform both the catalytic duties of proteins and the information-storage task of DNA.

So if only you could get an RNA molecule, or something like it, to step forward from the pre-biotic soup, life would be almost explained.

But no one has yet come up with any remotely plausible mechanism for RNA molecules to emerge naturally from pre-biotic chemicals. Indeed, the more work is done on this scenario, the less likely it seems.

It is amid all these promising but so far barren leads that Dr. Wächtershäuser proposed a fresh theoretical approach and backed it up with some interesting chemical tests. So far, his theory—that life must have started as an open metabolism, a cycle fed by some natural source of energy that synthesized a new molecule at each turn—is far from proven. But it is also an imaginative new approach to an outstanding problem, and all the more remarkable for having been made by someone who is a scientist only in his spare time.

—N.W.

Steven Weinberg

Physicist Ponders God, Truth and "Final Theory"

Dr. Steven Weinberg is perhaps the world's most authoritative proponent of the idea that physics is hurtling toward a "final theory," a
complete explanation of nature's particles and forces that will endure as
the bedrock of all science forevermore. He is also a powerful writer whose
prose can illuminate—and sting.

His withering essay on the dangers of utopian thought was promi-
nently featured in this month's *Atlantic Monthly*. The third volume of his
Quantum Theory of Fields, a weighty work on matter and energy at their
most fundamental levels, is soon to be released by Cambridge University

Press. And he recently received the Lewis Thomas Prize, awarded to the researcher who best embodies "the scientist as poet."

All of this combines two of his major passions: theoretical physics, which won him a Nobel Prize in 1979, and his often polemical writings on culture, religion, philosophy and, in particular, the history and politics of science.

At 66, he shows little sign of cutting back.

Dr. Weinberg, who grew up in the Bronx and is now a professor of physics and astronomy at the University of Texas at Austin, has little patience with the attempts by philosophers to explain how and why scientific theories are constructed, an activity that he regards as a squirrelly intrusion on working scientists. He dislikes any suggestions that the "truths" of science might to some degree be artificially constructed, and therefore subject to change by different human cultures at different times.

As for human spirituality, "I don't even know what it means," he said. He sees no redeeming value in religion and considers it nonsense.

But though his writing reveals him a sometimes hard-edged thinker who follows the evidence without regard to tradition or sentiment, face-to-face he is frankly romantic, deeply touched by music and poetry in ways he admits reason can never justify or explain. "I love grand opera," he abruptly confessed in an interview in his book-lined study. "I can't hear *La Bohème* without dissolving."

He enjoys hanging out with politicians, artists, writers and ranchers, and has friends outside the university orbit, in the heart of the Bible Belt. "There's a lot of good humor here," he said, adding that when he expounds his views on religion a common reaction goes something like, "There's old Steve. He will go on that way."

Dr. Weinberg was driving his red Camaro to work at the Massachusetts Institute of Technology in Cambridge in 1967 when he had an idea that changed physics. He realized that it might be possible to use a paradoxical-sounding idea about nature's fundamental order called broken symmetry to find the underlying unity in two of nature's four known forces or interactions.

The first, electromagnetism, involves the everyday forces applied by electric and magnetic fields, and the second, called the weak force, causes the radioactive decay of elements like radium and uranium. Electromagnetic forces are thousands of times as strong as the weak interaction, and they are transmitted by a particle with no mass, called the photon. As it turns out, the weak interaction is transmitted by very heavy particles called the W and Z.

But as he drove along in his Camaro, Dr. Weinberg saw that the same equations could describe the two interactions if a kind of energy, called a scalar field, permeated all of space. The field would in effect nudge the interactions in different directions, so that the underlying symmetry of the equations was broken, or hidden.

Physicists are still searching for direct evidence of that field—it should spawn something called the Higgs particle—but they have accepted Dr. Weinberg's idea, now called the electroweak theory. His paper explaining it became the most frequently cited paper in the recent history of particle physics.

"It was absolutely like lightning suddenly flashed," said Dr. Freeman Dyson, a physicist at the Institute for Advanced Study in Princeton. "It was immediately obvious that it was great."

In 1979, Dr. Abdus Salam, a physicist from Pakistan, and Dr. Sheldon L. Glashow, a classmate of Dr. Weinberg at the Bronx High School of Science, shared the Nobel Prize for the work.

The core of the electroweak theory went on to inspire a successful and experimentally verified theory of the strong force, which holds atomic nuclei together. The results were bundled together in what physicists call the Standard Model, a theory that includes all the known forces except gravity and amounts to a particle physicist's bible.

Dr. Edward Witten, a physicist at the Institute for Advanced Study, described the theoretical innovations at the heart of the electroweak theory as "the main lessons of elementary particle physics in the last half century."

Beyond his many research papers, Dr. Weinberg has also made a mark with writings that include scientific monographs like *Gravitation and Cosmology* and *The Quantum Theory of Fields,* popular books like *Dreams of a Final Theory* and *The First Three Minutes: A Modern View of the Origin of the Universe* and articles and essays on culture and science in publications like *The New York Review of Books, Scientific American* and *George* magazine.

Steven Weinberg was born on May 3, 1933, and grew up so close to Yankee Stadium that "the lights kept you up at night," he said. His mother and his paternal grandfather immigrated from Europe. Much of his mother's family in Germany died in the Holocaust. His father made a living as a court stenographer.

As a boy, he absorbed classical music on the radio and learned chemistry from a hand-me-down set, ChemCraft #5. Despite his proximity to the ballpark, he was not a sports fan. (He still prefers baseball to football because, he said, it is easier to tell who has the ball.)

At Bronx Science, where "it was considered very uncool not to have learned calculus on your own," he began to blossom as a physics student, particularly after he encountered a popular book by Sir James Jeans that mentioned quantum physics, and its mysterious equations and engrossing idea that nature was based on simple but powerful laws.

"There was something at the bottom that was much simpler than the appearance," he said. "That was the cutting edge of knowledge."

He attended Cornell on a scholarship, where he studied physics and fell in love with a fellow student, after hearing her sing in a student production. They married and today Louise Weinberg is a professor of law at the University of Texas.

After a "wonderful, romantic first year" in Copenhagen, where Steven Weinberg did research at what is now called the Niels Bohr Institute, the couple returned to the United States, where he earned his Ph.D. at Princeton and worked at a series of universities, including Columbia, the University of California at Berkeley, MIT and Harvard, where Louise Weinberg studied law. In 1980, she won an appointment at the law school at the University of Texas and in 1982, her husband followed her there.

It was there that he carried out the biggest battle of his professional career. He and other physicists believed they would need a gigantic particle accelerator to find the Higgs particle; eventually the government proposed building it near Waxahachie, Texas. But vigorous lobbying by scientists could not overcome Congressional dismay over delays and high costs for the multibillion-dollar Superconducting Supercollider, and Congress killed it in 1993. Dr. Weinberg attributed its death to overzealous cost-cutting by freshman Democrats.

Though the loss of the collider was a terrible blow to physicists in the United States—many believe leadership in the field has shifted now to Europe—Dr. Weinberg has happy memories of the battle, recalling with particular fondness an appearance on the *McNeil-Lehrer News Hour* with Senator Phil Gramm, a conservative Texas Republican.

"It was the only time Phil Gramm and I were aligned," said Dr. Weinberg, whose politics are sharply liberal.

Today, one of his major battles is with postmodernist thinkers and philosophers of science who maintain that scientific theories reflect not objective reality but social negotiations among scientists. In its rawest form, this philosophy would say that the theories of the most persuasive or politically powerful scientists become accepted fact.

Dr. Weinberg wrote of one such book on the subject, *Constructing Quarks,* by Dr. Andrew Pickering, that social negotiations in research are similar to the planning that mountain climbers might undertake together before tackling Mount Everest. But no one would think of writing a book called *Constructing Everest,* Dr. Weinberg said; once they had seen the mountain peak, most people would accept that it, like the elementary particles that leave their traces in particle detectors, had been shown to exist and had not been "constructed" by social agreement.

In general, Dr. Weinberg said, he believes that "half-baked philosophy has sometimes gotten in the way of doing science."

And then there are his pronouncements on religion and deism, including his much-quoted aphorism, "The more the universe seems comprehensible, the more it also seems pointless."

But in the seldom-cited passages that follow, Dr. Weinberg professes belief in his own kind of conviction, the idea that the scientific effort to uncover a complete theory of the universe is one of the things that can in itself add dignity and meaning to human existence.

As for conventional religion, though, his views are uncompromising: it is not only silly but damaging to human civilization. "The whole history of the last thousands of years has been a history of religious persecutions and wars, pogroms, jihads, crusades," he said. "I find it all very regrettable, to say the least."

Actually, Dr. Weinberg does occasionally entertain the possibility that there might be a God. While sitting in his study, with its striking view of Lake Austin, he imagined himself in the role of the biblical Abraham, whose faith God tested by commanding that he sacrifice his own son.

"Even if there is a God," Dr. Weinberg said, "how do you know that his moral judgments are the correct ones? Seems to me Abraham should have said, 'God, that's just not right.'"

—JAMES GLANZ
January 2000

SKEPTICAL TOURISTS IN PHYSICS LAND

Like a Fodor's guide to Manhattan, the science of physics is generally sold as the ultimate, indispensable, up-to-the-minute guide to reality. The parallel is far from exact: physicists cannot credibly tout their research as inexpensive and easy to use. But if there is one thing that the painstaking experiments and deep thinking of physics are meant to produce, it is theories that explain both the delights and the details of natural reality at its most fundamental level.

So it upsets physicists like Steven Weinberg that just as the field is enjoying unparalleled success, its relationship to reality is being questioned with a pointedness not seen for decades. Since the doubters are mostly nonphysicists—philosophers, historians, even the public in general—it is as if tourists from Keokuk had begun questioning just how unimpeachable the guidebooks are.

Logically, reality begins with elementary particles and the interactions that bind them to build the universe that our instruments measure, our eyes see and our hands touch. Physics, with its ambition of understanding these building blocks at the most basic level, therefore becomes a prime target for doubt.

Some motives for this skepticism can be found in the 20th century's scientific triumphs themselves, said Dr. George Levine, director of the Center for the Critical Analysis of Contemporary Culture at Rutgers University.

"I think it's provoked by the overwhelming authority that science has been able to accumulate," Dr. Levine said. "There's a sense that there are a few adepts, highly trained, who talk to each other, and there is no point of entry for anyone outside that area."

Driven by that sense of disenfranchisement, Dr. Levine said, anyone can scour the annals of philosophy for potential wedges to separate science from an immutable, objective reality that it seeks to explain. Philosophical pragmatists and empiricists, for example, hold that while a successful theory should describe experiments as accurately as possible, it has no necessary relation to ultimate reality, partly because one can never say whether new experiments will give new results and force changes in the theory.

More irksome to physicists is a school that finds its source in *The Structure of Scientific Revolutions,* a book first published in 1962 by the philosopher of science Thomas Kuhn. In this school, the form that theories take has as much to do with the politics and sociology of science as with its duty to describe reality.

The book seemed to show that science did not progress inevitably toward a better and better representation of reality, but instead that scientists moved from one "paradigm," or theoretical framework, to another based on social and psychological factors as well as experimental ones.

That view suggests that research in any paradigm brings science "closer to answering its own questions," said Dr. Steve Fuller, a sociology professor at the University of Warwick in England and author of *Thomas Kuhn: A Philosophical History for Our Times*. "But those questions are autonomous from the bigger question of how close they are getting to reality as such."

When pushed, physicists will often concede that there is room, at least linguistically, to wonder how absolute the correspondence is between successful theories and what might be regarded as ultimate reality.

Theories get revised in light of fresh evidence that they cannot explain. Sometimes the need for later revision is clear even as the theories are being devised: nearly all modern theories of matter and energy are called "effective field theories." That qualifier, "effective," means physicists already know that as the level of energy of things like particle collisions is pushed higher and higher, the theory gives wrong answers and cannot be trusted.

"In the past, we always required that our field theories would at least mathematically be consistent up to the highest possible energies," said Dr. Chris Quigg, a physicist at the Fermi National Accelerator Laboratory. "Now we admit the feeling of humility."

These revisions, however, do not rattle physicists, since the earlier theories almost always remain mathematically valid in some way or other. Effective field theories will still work at low energies, even after a fancier version is developed for higher energies.

Still, scientists, who depend heavily on public financing for their work, may wonder if doubts about the connection of physics and reality will someday keep the tourists, so to speak, from feeling an occasional need to visit the scientific paradise of ideas. Without those tourist dollars, the debate could become moot even as it remains unresolved.

—J.G.

Gary Larson

An Amateur of Biology Returns to His Easel

Gary Larson and his closest friends agree. If you want to understand the man—the comic genius, the author of the blackly buoyant and sorely missed *The Far Side* comic strip, and a cartoonist so revered among scientists that they have named a louse and a butterfly after him—then look at his work.

So let's skip the ahems, and start with a *Far Side* sampler, a few quick drill holes into Mr. Larson's sanctum delirium:

A scientist is standing on a podium, holding a duck. All the scientists in the audience also are holding ducks, save for one man, whose eyes are wide open in horror. The caption reads: "Suddenly, Professor Liebowitz realizes he has come to the seminar without his duck."

A woman is pushing a vacuum cleaner down a forest road and looking around nervously. The caption: "The woods were dark and foreboding, and Alice sensed that sinister eyes were watching her every step. Worst of all, she knew that Nature abhorred a vacuum."

A group of the damned are milling around the lobby to Hell, drinking coffee from an urn as though at a company reception. Devils surround them; flames lick through the door. One grumbles to another: "Oh man! The coffee's cold! They thought of everything!"

Gary Larson, too, has thought of everything, up, behind and athwart nature's mad phylogeny; and he has drawn everything, and he has put himself into the heads of all his creatures, including amoebas, which have no heads. And since he stopped doing his *Far Side* strip in 1995, he has left his tens of millions of fans in hell, where the coffee is always cold, and the bagels are always onion, because there is no Gary Larson.

Now, Mr. Larson is among us again, not as a syndicated cartoonist, but as a contemporary fabulist, a sort of green Gary Grimm who sides with the trolls and dryads. He has a new book out called *There's a Hair in My Dirt: A Worm's Story* (HarperCollins), a vividly illustrated narrative about a Father Worm, a Mother Worm, a sullen Son Worm, and Harriet—a blundering Panglossia with a tiara and blonde bouffant, who thinks nature is a Teletubby playground designed to enchant her. She serenades the cute, bulldozes over the creepy, and pays for her naïveté with the worst sort of hair day. And all around her real life goes on: a firefly "flashes" with a flick of his trench coat, a bear studies a "Field Guide to the Humans" ("Mushroomer: Usually seen in spring and summer. Shy, secretive, always looking down. Good eating.").

Mr. Larson, 47, has come east from his home in Seattle to do some very limited promotion, and to vacation with his wife, Toni Carmichael, 44, an anthropologist who helps run his multilegged enterprise, FarWorks. Over a long dinner and later in a jazz club—Mr. Larson is a passionate jazz lover and jazz guitarist—he talked about safari ants, Tarzan, gorillas in Uganda, Ivan the pet-store gorilla, whip scorpions, cows, ducks, Charles Addams and parasites. "I love parasites!" he said, over his mahi-mahi. "I can't get enough of them. Bring me some more schistosomiasis!"

He talked about his book, the possibility of his doing a feature film, and the animated video he had just finished. "It was quite a challenge to do," he said. "I didn't want any dialogue in it, just visuals, screams and grunts." He talked about whether he might go back into the business of daily cartooning. "I don't think so," he said. "Never say never, but there's a sense of 'been there, done that.' "

And though he hates having a fuss made over him and his fame—"that's the F-word to me," he says—he talked about himself, too.

Mr. Larson is a man of medium build and height—"Five foot 10 on a good day, in my shoes," he says—and he walks with a distinctive wind-blown posture. "It's my Groucho slant," he says. Much to this reporter's dismay, he has no features that can be compared to his creatures.

He draws a lot of cows, but his eyes are not tragically bovine, they're washed-out blue, as he puts it. No beetle brow, no beakish nose, no snaking neck. He looks like what he is, a boomer from the Pacific Northwest, born and raised in Tacoma, Washington. He wears wire-rimmed glasses, blue jeans, a simple button-down blue shirt and running shoes. His boldest fashion statement is a Swiss Army watch.

He bears a slight resemblance to Richard Dreyfuss, and, like the actor, he has a reedy voice. He is shy and friendly, comfortable and cynical. He laughs a lot, jokes about his thinning hair, and doesn't forget a word that's said over hours of conversation.

Mr. Larson has been a phenomenally successful cartoonist by any measure. When he retired from daily cartooning, his *Far Side* panel appeared in 1,900 newspapers. He has published 22 *Far Side* books, and all but one have been best-sellers. They have been translated into 17 languages and have sold 33 million copies worldwide. He has sold 45 million *Far Side* calendars and 110 million *Far Side* greeting cards. And almost everything he has done is funny.

Mr. Larson stands out as the darling of the scientific community. For years, his cartoons graced the bulletin boards, supply cabinets and incubators of, oh, 98.6 percent of all laboratories, here and abroad.

"His influence is pervasive," said Dr. Harold Varmus, director of the National Institutes of Health. "I can't tell you how many seminars I've been to that had a Gary Larson slide in them."

Scientists love him because he strips science to its pith, and he gets it right. May Berenbaum, an entomologist at the University of Illinois who runs the university's Insect Fear Film Festival, said: "He covered an extraordinary diversity of insect biology, and he was usually dead on the mark. For example, insect fecundity—the housefly with thousands of pictures of maggots in a billfold. Or cannibalism—one female mantid saying to the other, 'How dare you insinuate I would eat your husband?' People are not accustomed to looking at things through compound eyes."

Entomologists paid tribute to Mr. Larson by naming a species of butterfly from the Ecuadorian rain forest the *Serratoterga larsoni,* and a species of chewing louse found only on owls the *Strigiphilus garylarsoni.*

Mr. Larson's love of the swamp and all plasm within began in childhood. He and his only sibling, an older brother named Dan, spent many hours by the waters of Puget Sound at low tide, wading in their boots, swinging their nets. They caught grunt fish, octopus, salamanders, sea anemones. "We had this theory that all naturalists suffer from the 'oh please, oh please' syndrome," he said. "You're wading somewhere, and you see the biggest and most beautiful whatever. And all you can think, as you try to get up close, is, 'Oh please, oh please.' "

In their basement, they built teeming terrariums and even had a miniature desert ecosystem. Far from squawking about mess or stench, their father, a car mechanic and salesman, and their mother, a secretary, proudly invited the neighbors over for a tour.

Mr. Larson's taste for the nontraditional house guest continued into adulthood. For a while he bred Mexican king snakes, and kept a Burmese python until it grew 15 feet long. He has owned tarantulas and bird-eating spiders, African bullfrogs and carnivorous South American ornate horned frogs. Once, while he was washing a frog in the sink, the animal slipped down the garbage disposal. "He was O.K., but when I reached down to get him, he filled himself with air so I couldn't bring him up," Mr. Larson said. "I spent the longest time with my hand down the drain, waiting for him to relax, and at the same time not getting bitten. Frogs have teeth, you know." Mr. Larson said that, for environmental reasons, he no longer condones the rearing and keeping of exotic pets.

As a student at Washington State University, he started majoring in biology but changed course midway through college. "I didn't want to go to school for more than four years, and I didn't know what you did with a bachelor's in biology," he said, "so I switched over and got my degree in communications. I regret it now. It was one of the most idiotic things I ever did." Entomology, he said, "is my fantasy, the road not taken." His brother did major in biology and worked for a biological supply company before

THE FAR SIDE® OF SCIENCE
GARY LARSON

Science Meets Tabloid TV

THE FAR SIDE® OF SCIENCE

GARY LARSON

opening a plant nursery. He died four years ago, at the age of 46, from a sudden heart attack. "It was a profound loss for Gary," said Dan Reeder, a close friend of Mr. Larson who teaches high school mathematics in Seattle. "It was the only time I ever saw him really down."

Mr. Larson read plenty of comic books in his day—mostly *Tarzan*—and he always loved drawing, but he had no thought of becoming a cartoonist. After college he played the banjo professionally, in a duo called Tom and Gary. " 'As exciting as their name,' a friend of mine put it," Mr. Larson said. Eventually he traded the banjo for jazz guitar. In the mid-70's, he was on the verge of getting his dream gig, playing guitar for an established big band, but the bandleader ended up hiring somebody else. Crushed, Mr. Larson spent the weekend drawing cartoons. On Monday, he took them down to a small Seattle magazine, and the magazine bought them all. Two years later, in 1979, he signed a contract with *The San Francisco Chronicle* to do a cartoon panel six days a week; the publisher dubbed it *The Far Side*.

Mr. Larson said the relative ease with which he fell into cartooning explains why he became a cartoonist. "I don't think I ever had the stamina, or was thick-skinned enough, to go through a long process of trying to break in," he said. "I just started getting these motivations to keep going."

The great majority of his ideas for cartoons, Mr. Larson said, came straight from his head, and drew upon his early exposure to nature. He never farmed out his work to contractors, as highly successful cartoonists often do. He simply sat in his studio, and thought, and drew. "It's a strange, very isolated world," he said. "Time was amorphous for me while I was working. The only thing I knew was that the deadline was Saturday afternoon at 2 o'clock, because that was Federal Express's last pickup for Monday delivery."

He can't quite say how he came up with his ideas. Professor Liebowitz and his duck? "I've had those dreams of going somewhere in my underwear," he said. "I took that idea and married it to a serious scientific forum. But it's really all about a duck." Why are so many of his cartoons about cows? "I've always thought the word cow was funny," he said. "And cows are sort of tragic figures. Cows blur the line between tragedy and humor."

Wherever he plucked his ideas from, he struggled with perfecting each one—the graphics, the cadences of the captions, how many s's should be used for a leaking tire. "As his popularity mushroomed, the pressure built on him to perform even better," said his friend Mr. Reeder. "He felt he didn't have the luxury of producing even one cartoon that wasn't great."

Eventually, Mr. Larson got tired of feeling like there was always homework due. He was afraid of ending up, as he said at his retirement, in the Graveyard of Mediocre Cartoons. "It was an internal clock that told me, this is the time," he said. "I didn't feel that my identity was caught up in being a cartoonist, and that if it stopped I'd stop." He wanted to do videos, films, to play his guitar. He wanted to write *There's a Hair in My Dirt.*

And he has, and maybe this book is just the first in a series. As his Family Annelida so sweetly promises on the book's final page: see you soon.

—NATALIE ANGIER

April 1998

Index of Scientists